Library Services in Non-Library Spaces

SPEC Kits

Supporting Effective Library Management for Over Twenty Years

Committed to assisting research and academic libraries in the continuous improvement of management systems, OLMS has worked since 1970 to gather and disseminate the best practices for library needs. As part of its committment, OLMS maintains an active publications program best known for its SPEC Kits. Through the OLMS Collaborative Research / Writing Program, librarians work with ARL staff to design SPEC surveys and write publications. Originally established as an information source for ARL member libraries, the SPEC series has grown to serve the needs of the library community worldwide.

What are SPEC Kits?

Published six times per year, SPEC Kits contain the most valuable, up-to-date information on the latest issues of concern to libraries and librarians today. They are the result of a systematic survey of ARL member libraries on a particular topic related to current practice in the field. Each SPEC Kit contains an executive summary of the survey results (previously printed as the SPEC Flyer); survey questions with tallies and selected comments; the best representative documents from survey participants, such as policies, procedures, handbooks, guidelines, websites, records, brochures, and statements; and a selected reading list—both in print and online sources—containing the most current literature available on the topic for further study.

Subscribe to SPEC Kits

Subscribers tell us that the information contained in SPEC Kits is valuable to a variety of users, both inside and outside the library. SPEC Kit purchasers use the documentation found in SPEC Kits as a point of departure for research and problem solving because they lend immediate authority to proposals and set standards for designing programs or writing procedure statements. SPEC Kits also function as an important reference tool for library administrators, staff, students, and professionals in allied disciplines who may not have access to this kind of information.

SPEC Kits can be ordered directly from the ARL Publications Distribution Center. To order, call **(301) 362-8196**, fax **(301) 206-9789**, e-mail **pubs@arl.org**, or go to **http://www.arl.org/pubscat/**.

Information on SPEC Kits and other OLMS products and services can be found on the ARL Web site at **http://www.arl.org/olms/infosvcs.html**. The Web site for the SPEC survey program is **http://www.arl.org/spec/**. The executive summary or flyer for each kit after December 1993 can be accessed free of charge at the SPEC survey Web site.

SPEC Kit 285

Library Services in Non-Library Spaces

November 2004

Gordon Aamot

Head, Arts, Architecture, and Business Libraries

University of Washington

Steve Hiller

Head, Science Libraries/ Library Assessment Coordinator

University of Washington

Association of Research Libraries

Office of Leadership and Management Services

Series Editor: Lee Anne George

SPEC Kits are published by the

Association of Research Libraries
OFFICE OF LEADERSHIP AND MANAGEMENT SERVICES
21 Dupont Circle, NW, Suite 800
Washington, D.C. 20036-1118
(202) 296-2296 Fax (202) 872-0884
http://www.arl.org/olms/infosvcs.html
pubs@arl.org

ISSN 0160 3582

ISBN 1-59407-662-6

SPEC Kit 285

Library Services in Non-Library Spaces

November 2004

SURVEY RESULTS

REPRESENTATIVE DOCUMENTS

SELECTED RESOURCES

SURVEY RESULTS

EXECUTIVE SUMMARY

Introduction

During the past decade, many academic libraries have seen declines in gate counts, use of print material, and in-person reference transactions. Changes in scholarly publishing and information technologies have dramatically impacted the manner and the degree to which students, faculty, and other library users use library facilities. The balance between print and electronic resources is shifting towards the latter and fewer users need to visit the library to find the information they need. Given the growth of wireless connectivity, students don't need to come to the library to work individually or collaboratively. They can do that in a number of spaces ranging from computer centers to student unions.

While libraries provide a wide range of virtual resources and services to meet new user demands, surveys and other research reveal that users find the library information environment increasingly complex and difficult to navigate. Library users want assistance at time of need and in their space. They also value personal contact with library staff as a way to ask questions and resolve problems.

Research libraries and librarians continue to seek new strategies for providing services to library users, including expanding services beyond library facilities and offering in-person library services in non-library spaces. This survey presents library administrators with an overview of what types of library services are being offered in institutional non-library spaces, the nature of the spaces, what type of library staff are participating, and how they are being marketed and funded. The scope of the survey focuses on regularly scheduled library services in non-library spaces and excludes course instruction and traditional library instruction services delivered in departmental classrooms.

Seventy-five of the 123 ARL member libraries (61%) responded to the survey. Forty-one of these (55% of the respondents and 33% of ARL member libraries) have offered or are offering scheduled, in-person services in academic departments or other institutional spaces outside of the library.

Background

A review of the literature shows that the provision of in-person library services outside the library has been somewhat limited, but it is not a new phenomenon. This is confirmed by the survey results: 20 of the 41 libraries (49%) that offer some type of service have done so for five or more years. However, the fluidity of this practice is demonstrated by two groups of respondents. Of the 20 libraries that began service more than five years ago, 12 (60%) have expanded service to additional locations within the last five years. At the same time, 11 of the 41 libraries (27%) have begun offering such services only within the last two years.

Regardless of the longevity of these services, first contact has typically been with faculty. The early adopters most often moved out to academic departments, hospitals or other clinical settings, and computer labs. A significant number of more recently begun services follow this pattern, though the institutions that have initiated services within the last two years seem to be focusing more on where students gather, such as residence halls, study halls, and student unions. Of those services located in departmental spaces, the subject areas served cover a broad range of academic endeavor, led by the social sciences and descending through the disciplines to the fine arts and area studies.

One of the most interesting things revealed in the survey are the major factors involved in the libraries' decisions to initiate the services. Eighty percent of the respondents indicated that individual librarian initiative was a major factor in starting the services. A changing service philosophy (65%) and user demand (58%) were the next highest reasons. Only a third of the respondents indicated that a decision by the library administration was a major consideration in initiating the services and only one indicated that the closing or merging of a branch library drove the decision. The high percentage of services in non-library spaces initiated by individual librarians indicates that such services are entrepreneurial and opportunistic in nature.

Seventeen libraries (42%) reported that a service in a non-library space had been started but, at some point, discontinued. Space constraints or reassignment of the space by the department, low use of the service or a prevalence of non-library-related questions, and changes in staff assignments or the departure of the library staff member who initially instigated the program were the reasons most often reported.

Services Offered

The survey asked respondents which library services are provided in which locations. The locations—in order of frequency used—are offices or workspace in departments, hospitals or other clinical settings, computer labs, residence halls, study halls or commons, career centers, student unions, writing centers, and research labs. A variety of other locations were also mentioned, including community centers, continuing education centers, and a student athlete study center. There is some understandable variation in service offerings across the locations. While provision of reference service and individual or group consultation ranks high across all locations—especially in departmental offices—workshops and demonstrations rank highest in computer labs, career centers, and hospitals or other clinical settings. Technical assistance is most often available in departmental spaces, computer labs, research labs, and writing centers.

There appears to be no consistent name for library services in non-library spaces. Slightly more than a quarter of the respondents use no particular name; the terms most often used are outreach (29%) and office hours (23%). Some more descriptive names include Librarian On-Site, Librarian on Location, Librarian in the Atrium, The Librarian Is In, Mobile Librarian Services, Field Librarian Program, MyLibrarian Project, House Calls, and Clinical Medical Librarian. The lack of a standard nomenclature reflects the entrepreneurial, ad hoc nature of many of these out-of-library services.

Although these initiatives are largely driven by individual librarians, 66% of the survey respondents report that the user community was involved in planning for these services. This input included negotiations on space, equipment, and service hours; technical assistance; and help with publicity.

Staffing and Other Resources

Respondents were asked to indicate both the number of librarians and other staff members participating in these services and the total number of hours per week that the service is offered by a librarian or other staff. The responses were broken out across all the service locations. Librarians provide service in all locations. The location with the highest number of participating librarians is office or workspace in an academic department (28 responses). Sixteen of the libraries (57%) that provide service there have just one or two librarians involved, but nine (32%) have between three and nine participating librarians, and three (11%) have twelve to fifteen. The total number of weekly service hours provided in this location ranges from two to forty, with 56% of respondents staffing up to seven hours per week and the rest staffing eight or more hours per week. Hospitals or other clinical settings, computer labs, and residence halls are the next most popular locations staffed by librarians but typically there are only one or two librarians involved. Service in these locations is available fewer than eight hours a week in all but a couple of cases. Other library staff—which includes student assistants—provide service in fewer locations. They most often staff services in residence halls, departmental spaces, student unions, and computer labs and typically work less than eight hours a week.

In addition to providing services in non-library locations, participating library staff members have a wide range of other responsibilities in the institutions in which they work. In the case of librarians, the vast majority carries a standard load of public service duties. These include reference / information services, instruction, collection development duties, and departmental liaison duties. The regular duties of other staff who participate in these programs primarily involve reference / information and instruction.

Not surprisingly, there is the same wide variety of special names or titles for participating staff as seen above in names for these service programs. Some are, indeed, specialized and emphasize the non-library space aspect. These include Field Librarian, Mobile Librarian, College Librarian, and Outreach Librarian. Others are more traditional, including Information and Education Services Librarian, Subject Librarian, and Liaison Librarian. When on duty in a non-library space, participating staff use a variety of means to identify themselves to users, most commonly with appropriate signage (13 responses or 48%). Some other methods include wearing a special identification badge, having flyers or brochures on hand, or simply being present at an assigned desk or office.

Library services in non-library spaces are offered on weekdays during the day by all but two of the responding libraries. Weekday hours are most common for services provided in offices or work spaces in academic departments (28 responses or 76%). Services provided in computer labs and hospitals have similar daytime patterns. Services located in the more student-focused spaces—such as residence halls, study halls, career centers, student unions, and writing centers—also have service hours on weekday evenings and Saturdays. A hospital is the only non-library space where some services are also provided on Sundays.

Staff provide library services in a variety of workspaces in non-library locations. The only spaces where librarians work from separate (42%) or shared (22%) offices in any significant number are in academic department spaces. Librarians in academic space also work from an assigned desk in an open area (17%), in unassigned space as available (14%), or "roam" the space and approach users one-by-one (19%). A few respondents have assigned space in other locations, but most roam the space to provide assistance.

Library staff use a variety of equipment to provide services, including desktop computers, laptop computers, printers, phones, and personal digital assistants. Six respondents also report occasionally using projection equipment. Office, clinical, and lab spaces are most likely to have a full compliment of equipment. Staff in student-oriented spaces rely more heavily on laptops. Printers and phones are virtually nonexistent outside of office-type spaces.

Funding

The survey asked how services in non-library locations are funded. All respondents report that services are funded primarily from the library budget. Three of the respondents indicated they received new funding to provide the services. About half receive some funding from a scattering of other sources. It is clear, though, that respondents consider departments' contribution of space and equipment shared funding for the service. Funding by non-library sources is used for equipment, librarian and other library staff time, and promoting the services.

Service Promotion

The most common means of promoting library services in non-library spaces has been e-mail announcements (91%), flyers or posters (71%), library orientations and instruction sessions (68%), and library Web pages (68%). Other publicity venues include departmental Web pages, library newsletters, newspapers, phone calls, and word of mouth. No matter what type of space, libraries have used a variety of methods to publicize the service. Where user communities are more defined, such as academic departments, hospitals, computer labs, and writing centers, e-mail has played a larger role in promoting the service. In spaces that serve broader, more interdisciplinary groups of users, posted signs, brochures, and Web pages have played greater roles in the marketing mix.

Service Evaluation and Assessment

The majority of responding libraries have not formally evaluated these services and 20% have not done any evaluation. Most rely on informal feedback about the services and observation for evaluation. Among those that have done some type of structured evaluation, surveys are the method most frequently employed (34%).

Opportunities and Challenges

Respondents were asked to list up to three benefits and three challenges of offering library services in non-library spaces. Several themes emerged from the comments. One benefit is increased visibility for the participating librarians and the library by providing services on users' turf. Another benefit is the user convenience of providing needed services on users' own terms. A third benefit is the opportunity to establish relationships and connect with faculty and students in a way that can't be done in a traditional library setting.

Challenges mentioned include the process of negotiating and procuring access to scarce departmental space. Staffing and keeping the service a priority for the libraries is also a challenge. These services can be relatively labor intensive and require committed library staff and administrations to make them successful. If key participants leave for other jobs or are asked to take on additional responsibilities elsewhere in the library, non-library service programs can fall by the wayside. A related challenge is the question of scalability. How can a library provide this service for every department? Or, should it try to provide the service for every department? Lastly, respondents said that marketing and publicizing the programs are an ongoing challenge.

Conclusion

At least one-third of ARL member libraries are providing, or have provided, some form of scheduled library services outside of the library. As information technology evolves and "any time, any place" access to resources becomes more ubiquitous, few users need to come to the library to find the information they need to do their work. Research libraries and librarians will continue to seek new strategies for providing services to library users, including working with them on their own turf.

There are several conclusions about the nature of library services in non-library spaces one may draw from the survey. To flourish, these services appear to require a delicate balance of entrepreneurial and dedicated staff able to absorb the additional duties without additional funding, the ability to take advantage of local opportunities, access to appropriate non-library workspace and technology, and interested users. Individual librarian initiative was reported as the major factor in instigating new services; the reassignment or resignation of key staff members was also noted as a major factor in decisions to discontinue services.

With a few exceptions, library services in non-library spaces are not fully integrated into public service programs. Forty-two percent of the libraries surveyed report that services had been started and then discontinued, largely because the conditions listed above didn't exist. The wide variety of names given for the services and the librarians who provide them also indicates the fluid nature of the services. Lastly, relatively few libraries report that they have formally assessed these services. Given the highly entrepreneurial nature of the enterprise and scaling issues, sustainability is a legitimate concern.

This begs the question, "How does a library determine if the service is a success?" Should success be measured solely in quantitative terms by the number of questions answered, staff members and departments involved, and cost effectiveness of the service? Or is a more useful indicator for libraries the indirect benefits accrued along the way, such as increased goodwill, heightened visibility for the library, and the chance to build stronger relationships between librarians and library users? Both approaches are valid and the answer is most likely some combination and the precise balance between the two can and should shift over time, based on local conditions. This is clearly an area where many libraries have been willing to take risks and experiment with a different service model. At the same time, institutions have been willing to step back and discontinue the service if experience shows it to not be worth the investment of staff time and resources.

While the quantitative data gathered by the survey is informative, the many comments written by the respondents are especially important in establishing the context for understanding both the nature of the services and the challenges that libraries providing these services face. Although numerous challenges were noted, they seem to be outweighed by the opportunities and benefits—both direct and indirect—that come from reaching out to students, faculty, staff, and other library users in their own spaces and building personal connections between libraries and library users.

SURVEY QUESTIONS and RESPONSES

The SPEC survey on Library Services in Non-Library Spaces was designed by Gordon Aamot, Head, Arts, Architecture, and Business Libraries, and Steve Hiller, Head, Science Libraries / Library Assessment Coordinator, University of Washington. These results are based on data submitted by 75 of the 123 ARL member libraries (61%) by the deadline of June 18, 2004. The survey's introductory text and questions are reproduced below, followed by the response data and selected comments from the respondents.

Twenty years ago students and faculty came to the library to use collections, services, and as a place to do their work. For many students the library was their primary workplace, providing physical access to course reserves and a quiet space to write or study. Where once users had to come to the library for collections, reserves, renewals, and reference, today many of these can be accessed anywhere students and faculty can connect to the campus network. Students can find alternate work places in such locations as computer centers, learning areas, student unions, and other friendly gathering spots. Two other factors have emerged recently that also impact where library services are provided—the trend toward branch library closures or mergers and the declining costs of laptop computers and wireless technology.

Given these changes in access to and delivery of information, as well as the emergence of different types of work spaces for students, many libraries have seen declines in gate counts, use of print material, and in-person reference transactions. While libraries provide a wide range of virtual resources and services virtually through library Web sites, surveys and other research reveal that users find the library information environment increasingly complex and difficult to navigate. Library users want assistance at time of need and in their space. They also value personal contact with library staff as a way to ask questions and resolve problems.

This survey is intended to gather information to provide library administrators with an overview of what types of library services are being offered in institutional non-library spaces, the nature of the spaces, what type of library staff are participating, and how they are being marketed and funded. The scope of the survey focuses on regularly scheduled library services in non-library spaces and excludes course instruction and traditional library instruction services delivered in departmental classrooms.

The survey is organized into two sections. The first part asks the what, where, who, and how questions about services being currently offered. The second part inquires about future plans.

BACKGROUND

1. Has your library offered any scheduled in-person services in academic departments or in other institutional spaces outside the library? Example of such services included scheduled office hours, reference, consultation, workshops, demonstrations, technical assistance, and clinical support. Do not include course instruction or other ad hoc activities. (n=65)

Yes	41	55%	Please complete the survey.
No	34	45%	Please skip to question 19.

If yes, please indicate how long ago service began in each location where service has been offered. Check all that apply. (n=40)

	< 1 year	1–2 years	3–5 years	> 5 years	Don't Know	Cumulative Responses
Office or work space in department	6	5	6	11	2	30
Hospital or other clinical	2	1	1	8	1	13
Computer lab	4	3	—	6	—	13
Residence hall	4	4	2	1	—	11
Study hall or commons	4	—	1	—	2	7
Career center	1	3	2	—	1	7
Student union	4	—	—	1	1	6
Writing center	1	1	2	—	—	4
Research lab	—	1	1	1	—	3
Other location, please specify	—	2	1	7	—	10
Cumulative Responses	15	14	13	21	3	

Other locations include:

Faculty offices, community centers (1–2 years)

Student Athlete Study Center (3–5years)

Lobbies of campus buildings, conference rooms, individual faculty offices, continuing education centers off-grounds, and department lounges (> 5 years)

Have any library services in non-library spaces been started but for some reason discontinued? (n=41)

Yes	17	42%
No	24	59%

If yes, please indicate which service(s) and the reason(s) it was dropped.

Selected Comments from Respondents

"Changing technology and space constraints forced the closing of the LRCs. Librarian who was in the natural sciences museum retired and position was repurposed."

"We had reference hours in an office in a humanities class building one year ago, but there was not enough traffic to warrant keeping that service."

"Office was needed for teaching faculty member."

"Biomedical library discontinued efforts in the student union because it was not a suitable IT environment."

"One of the programs ended when the education librarian took on new managerial responsibilities and was no longer liaison to the program."

"Librarian presence in the College of Human Medicine student computing lab: no use of his assistance."

"We held office hours in the writing center five years ago but discontinued because of organizational changes. We started a new collaboration last year but now the writing tutors hold office hours in two of our libraries."

"In academic department 'office hours' were discontinued following the departure of the single librarian who had the initiative and rapport to establish it with one department."

"Librarians piloted a reference service called 'Librarians Without Borders' in a student center. It was discontinued because the pilot period ended and participating librarians noted that further study was needed because not many questions were truly reference."

"Satellite reference location in the student union was discontinued after two years due to staffing changes, lack of traffic, and unsatisfactory location. In the mid-1990s, the science and engineering library offered the 'Internet Rx' service. This service involved liaison librarians and technical support staff going to faculty offices to help faculty connect to and use the Internet effectively. The service was discontinued when Internet connections became routine in faculty offices."

"College of Business Career Center: 1) students confused business reference and career center services; 2) students preferred librarian to constantly be in one location; 3) students preferred virtual reference; and 4) unstable wireless connections. Student Union: best location unavailable."

"Funding ceased."

"Reference service in the fine arts department discontinued due to unavailability of Internet connection and access to library Web pages; low patron turnout; space used by librarian not always available."

2. If services are provided in departmental spaces, which subject areas are involved? Check all that apply. (n=38)

Social Sciences	25	66%
Humanities	17	45%
Business	17	45%
Health Sciences	15	39%
Sciences	13	34%
Engineering/Computer Science	12	32%
Fine Arts	9	24%
Area Studies	6	16%
Other, please specify	3	8%

Please list the names of the specific departments below.

Social Sciences

Anthropology (3)

Communications

Criminal Justice

Economics (2)

Education (9)

Government

History (3)

Liberal Arts and Human Resources

Political Science (3)

Psychology (2)

Sociology (2)

Humanities

Classics

English (4)

Film and TV Studies

German and Scandinavian Studies

Philosophy

Religious Studies

Rhetoric

Business (10)

Health Sciences

Dentistry (2)

Medicine (3)

Nursing (4)

Occupational Therapy

Pharmacy (2)

Psychiatry

Physiotherapy

Public Health

Veterinary Medicine

Sciences

Agriculture and Life Science

Chemistry

Geology (2)

Mathematics

Natural Resources (2)

Physics

Science

Engineering/Computer Science

Chemical and Materials Engineering

Nanotechnology

Mechanical Engineering

Civil Engineering (2)

Information Science and Technology

Electrical and Computer Engineering

Mining and Petroleum Engineering

Fine Arts

Architecture (5)

Fine Arts

Music (2)

Area/Ethnic/Women's Studies

African American Studies

Black Studies

Jewish Studies

Latin American Studies

Scandinavian Studies

Women's Studies

Other

Law

3. **Please indicate the major factors in the library's decision to initiate these services.** Check all that apply. (n=40)

Individual librarian initiative	32	80%
Changing service philosophy	26	65%
User demand	23	58%
Decision by library administration	13	33%

Opportunity for space in department	10	25%
Non-library faculty or staff initiative	9	23%
Recommendation by library staff/committee	5	13%
Distance education program	2	5%
Closing or merging of branch library	1	3%
Other, please explain	6	15%

Selected Comments from Respondents

"For some of the branch libraries, low attendance in classes and lower library gate count."

"More convenient, better technology-equipped teaching space than library."

"The library is part of official programming that is allowed to go into housing areas."

"Extension of existing service philosophy, not really a change of service philosophy."

"Collaborative initiatives with student services units."

4. If the closing or merging of a branch library was a factor in the decision, please indicate the reasons for the closure or merger. Check all that apply. (n=21)

Changing library service philosophy	1	5%
Changing branch library user patterns	1	5%
Availability of electronic information at desktops	1	5%
Another unit needed the space	—	—
Library budget constraints	—	—
University de-prioritization of academic program	—	—
Other, please explain	—	—
Not applicable	20	95%

SERVICES

5. Which library services are provided in the locations listed below? Check all that apply. (n=40)

Office or Work Space in Department (n=30)	
Individual consultation	30
Reference	27
Group consultation	21
Workshops/demos	20
Technical assistance	18
Clinical support	4
Other service, please specify	—

Hospital or Other Clinical (n=13)	
Workshops/demos	11
Individual consultation	9
Clinical support	8
Group consultation	7
Technical assistance	7
Reference	6
Other service, please specify	—

Computer Lab (n=12)	
Workshops/demos	7
Reference	6
Individual consultation	6
Group consultation	5
Technical assistance	5
Clinical support	—
Other service, please specify	—

Residence Hall (n=9)

Individual consultation	7
Reference	6
Group consultation	4
Workshops/demos	4
Technical assistance	1
Clinical support	—
Other service, please specify	1

Whatever is needed by students.

Study Hall or Commons (n=6)

Reference	4
Individual consultation	4
Group consultation	2
Technical assistance	2
Workshops/demos	1
Clinical support	—
Other service, please specify	—

Career Center (n=5)

Workshops/demos	4
Individual consultation	4
Group consultation	3
Reference	3
Technical assistance	1
Clinical support	—
Other service, please specify	—

Student Union (n=4)

Reference	2
Individual consultation	2
Workshops/demos	2
Group consultation	—
Technical assistance	—
Clinical support	—
Other service, please specify	—

Writing Center (n=3)

Reference	3
Technical assistance	3
Individual consultation	3
Group consultation	2
Workshops/demos	2
Clinical support	—
Other service, please specify	—

Research Lab (n=3)

Individual consultation	3
Workshops/demos	3
Technical assistance	3
Reference	2
Group consultation	2
Clinical support	1
Other service, please specify	—

Other Location, please specify (n=10)

Individual consultation	7
Reference	5
Workshops/demos	5
Technical assistance	4
Group consultation	3
Clinical support	—
Other service, please specify	2

Other Location	Service(s) Offered
Business school	Reference
Student athlete study center	Reference, individual consultations
Conference rooms	Individual and group consultations, workshops
Campus classrooms	Individual and group consultations, workshops, technical assistance
Continuing education centers	Reference, individual consultations, workshops, technical assistance, arrange for other meetings and provide support
Department lounges	Reference, individual and group consultations, technical assistance
Community centers	Internet access

6. What does your library call these services? (n=35)

Outreach (10 responses)

No particular name (9 responses)

Office Hours (8 responses)

Librarian On-Site (2 responses)

Librarian on Location

Librarian in the Atrium

The Librarian Is In

Mobile Librarian Services

Field Librarian Program

MyLibrarian Project

Gateway Library Services

UGL Union (Undergraduate Library in the student Union)

House Calls

Clinical Medical Librarian

VAULT (Virtually Anywhere Unlimited Library Technology)

Distance Education Support Services

7. Was the user community outside of the library involved in planning for these services?

Yes	25	66%
No	13	34%

If yes, please briefly explain their role and the process used.

Selected Comments from Respondents

"Dean and the department encouraged librarian to attend problem-solving learning sessions and to consult with individuals in their offices/areas."

"The Faculty of Engineering provided space for the service and consulted with the library on the optimum hours for the service in that space to reach the most students during the week."

"Discussion about the scope of the service, the time, frequency, and the nature of the services."

"The Graduate School of Management requested that we maintain office hours during peak student activity times."

"Administrative staff of college-designated space designed office and specified technology equipment."

"Involved the Composition/Writing Center and Academic Computing in developing Student Center for Research, Writing, and Instructional Technology."

"Faculty assisted with site selection, hours of service, and publicity."

"Each department was contact by their subject specialist and offered the chance to have a librarian do outreach in their area. They negotiated the hours and offered space."

"Helped set goals, specify location, determine service hours, and create Web site."

"Career center at the graduate management school conceived the idea of a librarian holding office hours every Tuesday night. The career center did all of the promotion and provided the space and equipment."

"RAs would indicate the need for librarians to come; they would often arrange for pizza and hold the session during a dorm floor meeting."

"College deans and faculty were consulted. Initially started with College of Business and what was then the College of Education. Two years later evolved into a program for each college."

"New programming for inclusion in residence hall educational programs was designed in collaboration with student services."

"In case of psychology and sociology, hours were selected based on the recommendation of the chairs. Psychology department support staff was involved to make sure that the office space was reserved for the librarian and that the computer used was connected to the department's network and printing facilities were provided. Information Services and Technology Department provided the technical assistance to make library resources available remotely to the librarian. (Many of the databases at that time were on library network and not available outside of the library.) In case of fine arts, informal discussion with a faculty member led to service offered in the department."

"Departments were instrumental in providing adequate space for librarians to 'set up shop.' Worked with academic programmers in residences to develop relevant programming."

8. Please indicate how many and what categories of library staff provide services at each location. Indicate how many total hours, on average, each category of staff works at each location during an average week. Complete all that apply. Note: "Other library staff" includes student assistants. (n=37)

Office or Work Space in Department (n=28)

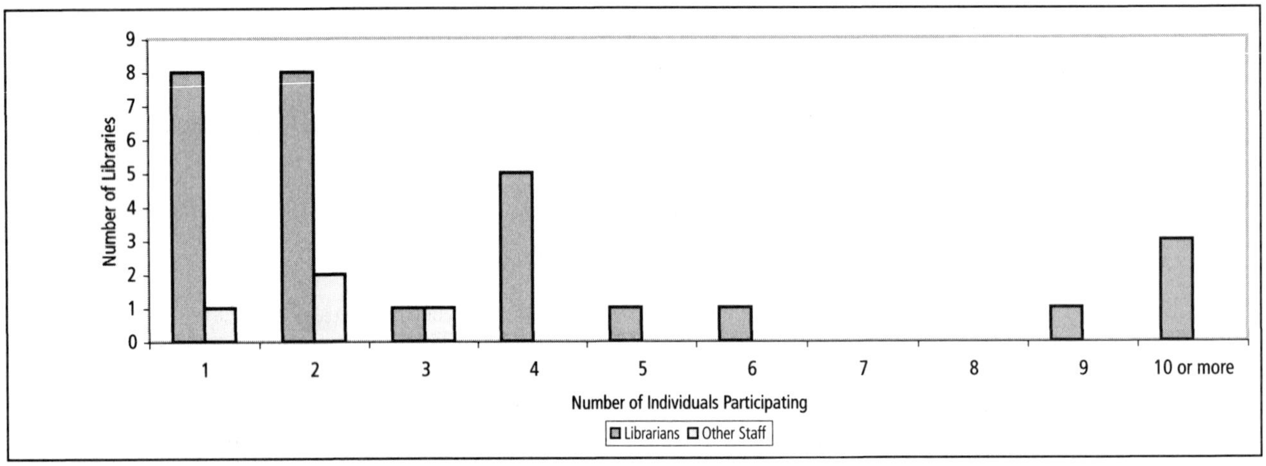

Participating	1	2	3	4	5	6	7	8	9	10 or more
Librarians (n=28)	8	8	1	5	1	1	—	—	1	3
Other Staff (n=4)	1	2	1	—	—	—	—	—	—	—

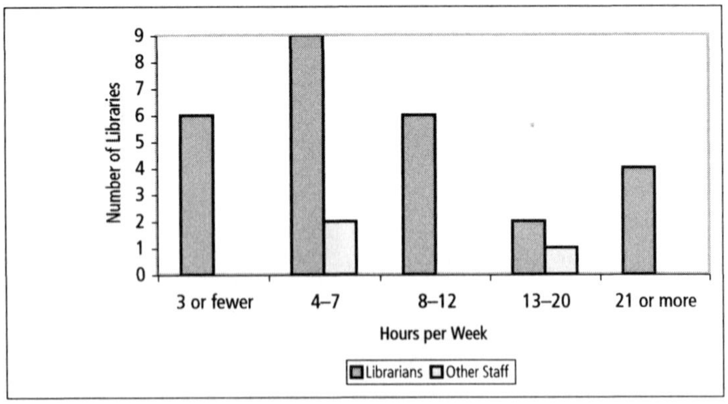

Hours	3 or fewer	4—7	8—12	13—20	21 or more
Librarians (n=27)	6	9	6	2	4
Other Staff (n=3)	—	2	—	1	—

Hospital or Other Clinical (n=11)

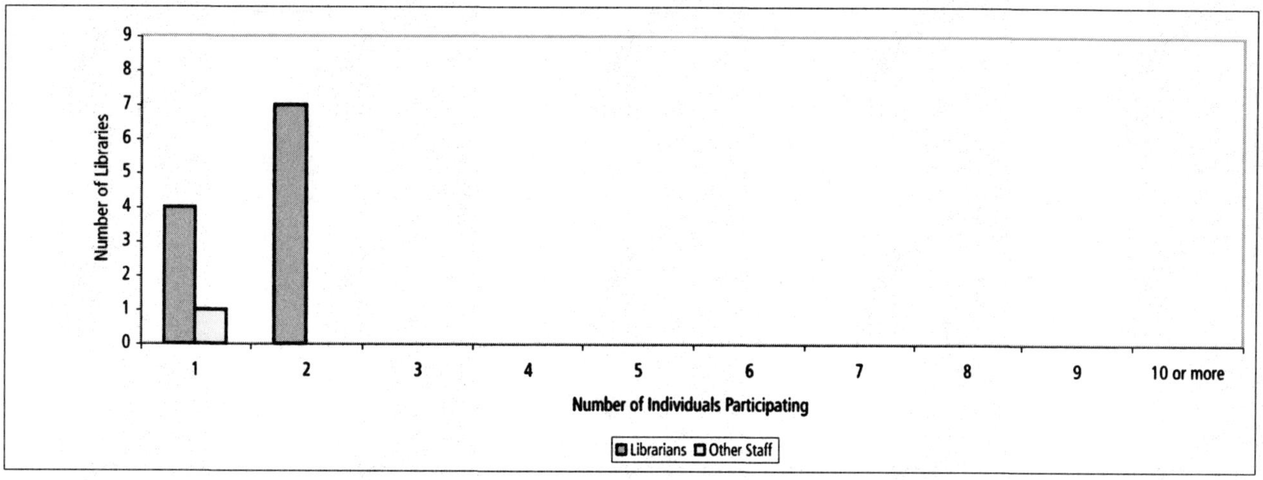

Participating	1	2	3	4	5	6	7	8	9	10 or more
Librarians (n=11)	4	7	—	—	—	—	—	—	—	—
Other Staff (n=1)	1	—	—	—	—	—	—	—	—	—

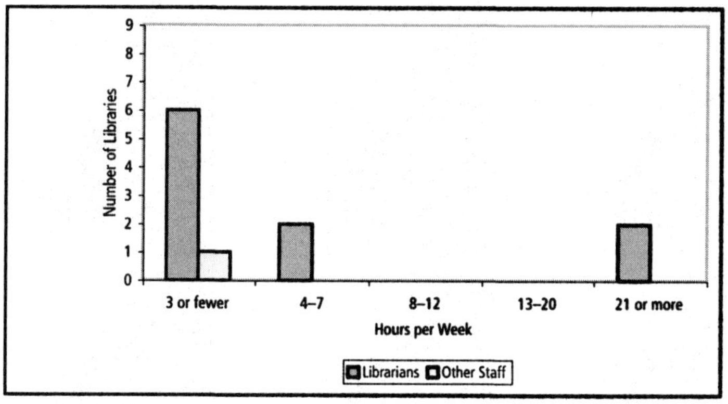

Hours	3 or fewer	4—7	8—12	13—20	21 or more
Librarians (n=10)	6	2	—	—	2
Other Staff (n=1)	1	—	—	—	—

Computer Lab (n=10)

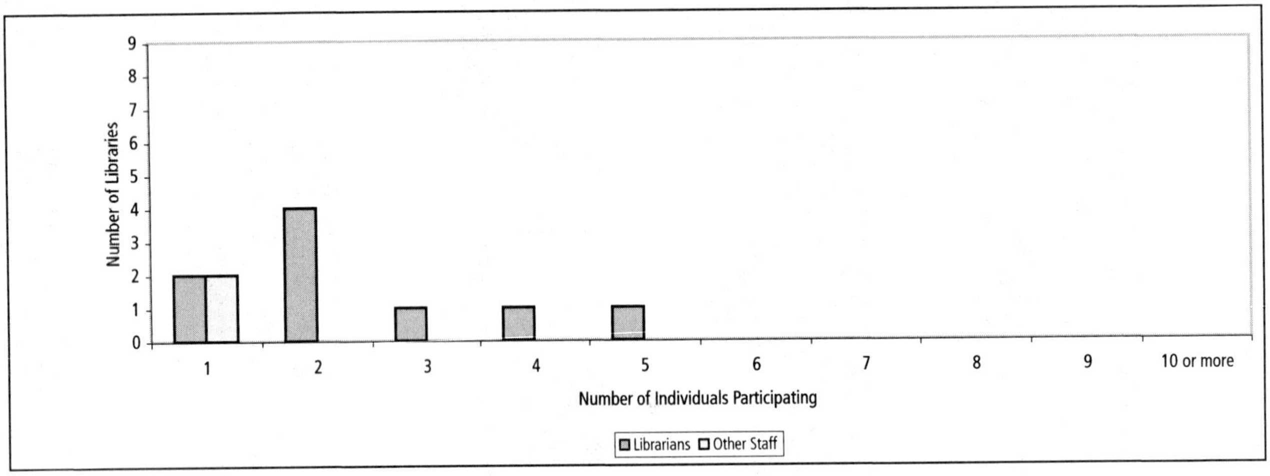

Participating	1	2	3	4	5	6	7	8	9	10 or more
Librarians (n=9)	2	4	1	1	1	—	—	—	—	—
Other Staff (n=2)	2	—	—	—	—	—	—	—	—	—

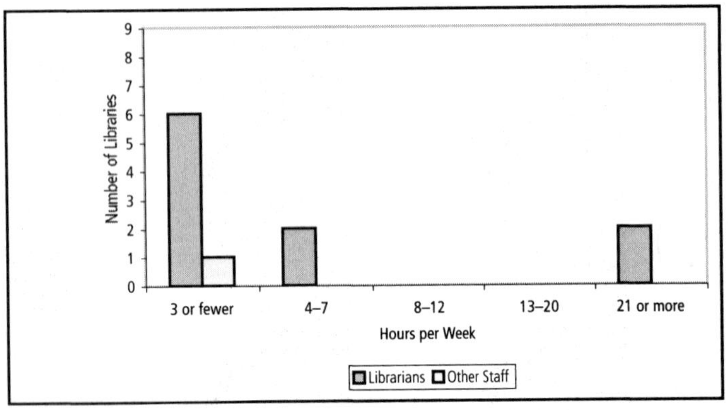

Hours	3 or fewer	4—7	8—12	13—20	21 or more
Librarians (n=10)	6	2	—	—	2
Other Staff (n=1)	1	—	—	—	—

Residence Hall (n=9)

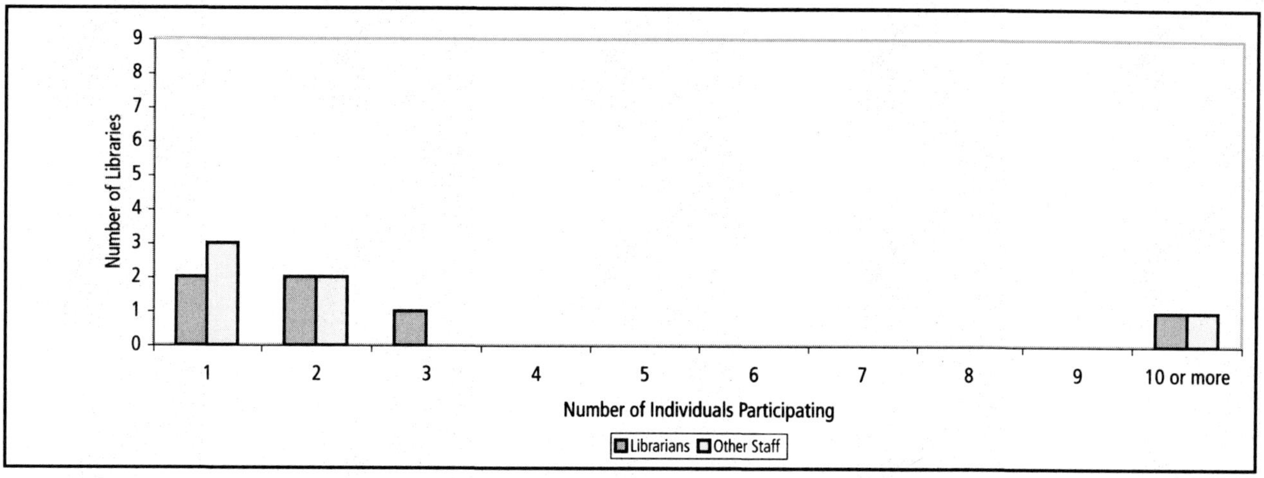

Participating	1	2	3	4	5	6	7	8	9	10 or more
Librarians (n=6)	2	2	1	—	—	—	—	—	—	1
Other Staff (n=6)	3	2	—	—	—	—	—	—	—	1

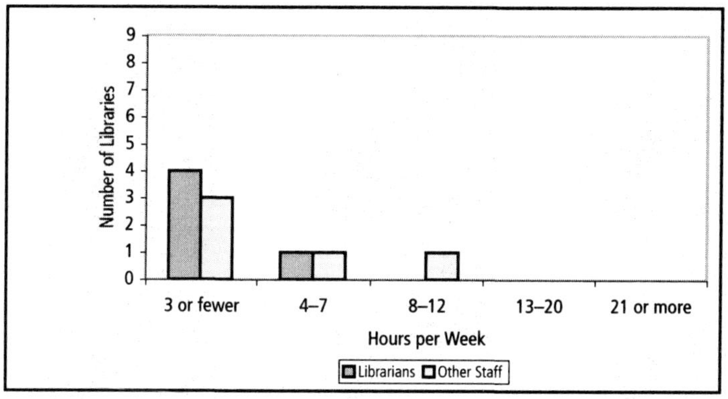

Hours	3 or fewer	4—7	8—12	13—20	21 or more
Librarians (n=5)	4	1	—	—	—
Other Staff (n=5)	3	1	1	—	—

Career Center (n=5)

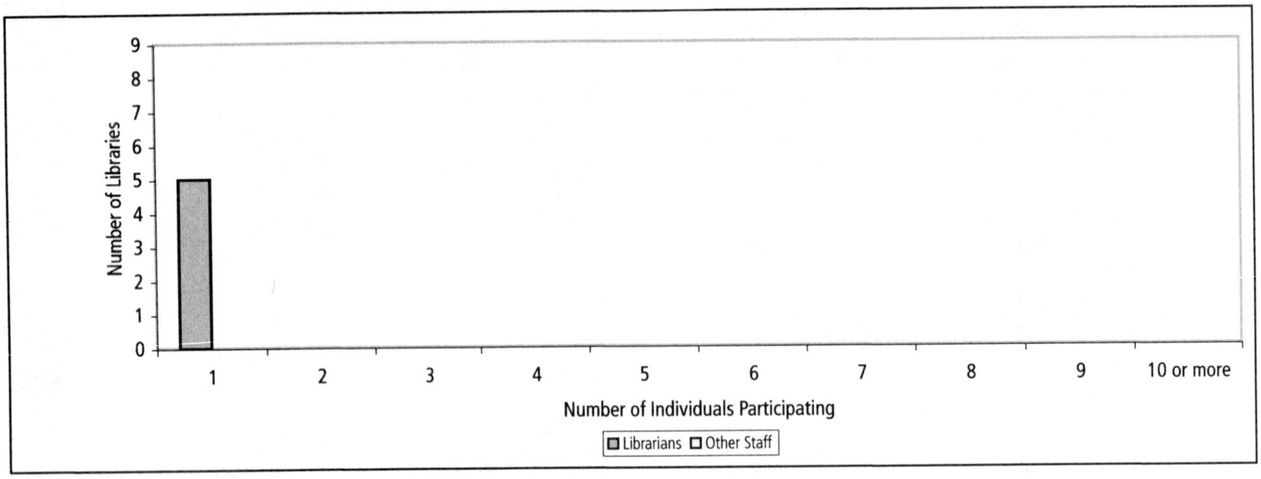

Participating	1	2	3	4	5	6	7	8	9	10 or more
Librarians (n=5)	5	—	—	—	—	—	—	—	—	—
Other Staff (n=0)	—	—	—	—	—	—	—	—	—	—

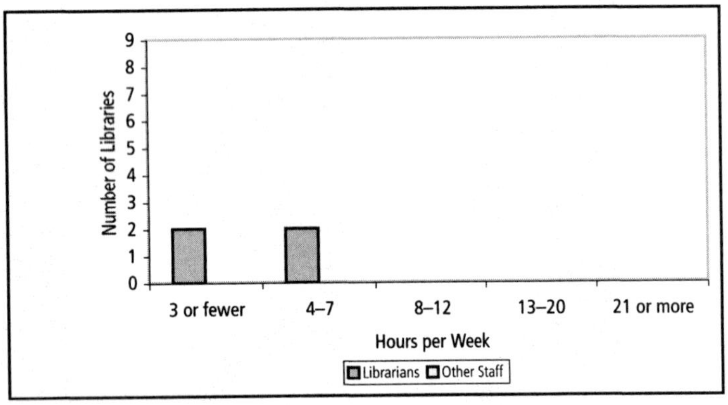

Hours	3 or fewer	4—7	8—12	13—20	21 or more
Librarians (n=4)	2	2	—	—	—
Other Staff (n=0)	—	—	—	—	—

Student Union (n=4)

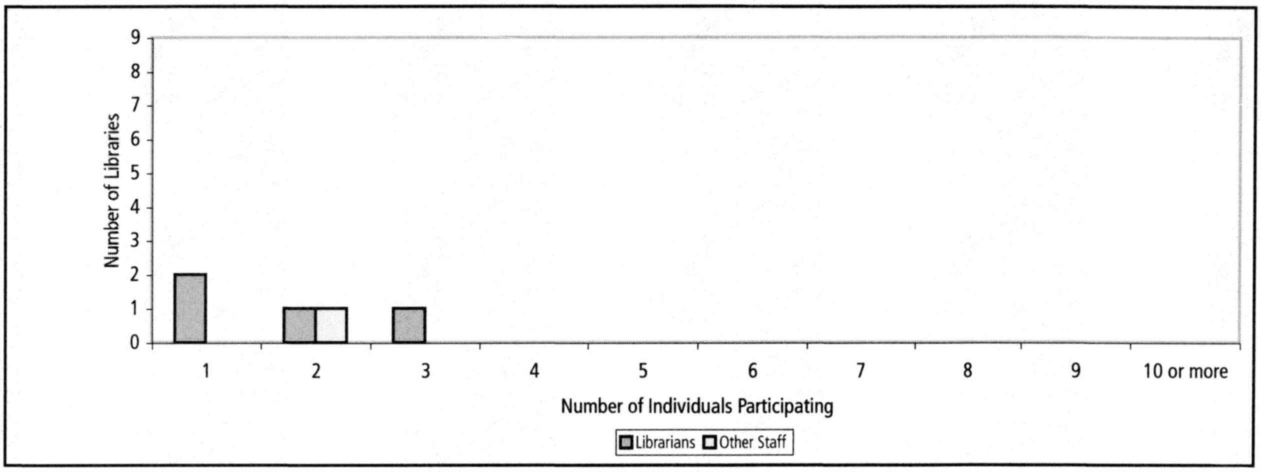

Participating	1	2	3	4	5	6	7	8	9	10 or more
Librarians (n=4)	2	1	1	—	—	—	—	—	—	—
Other Staff (n=1)	—	1	—	—	—	—	—	—	—	—

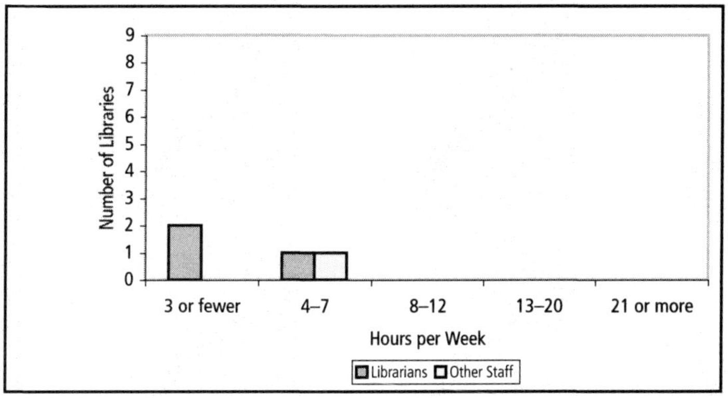

Hours	3 or fewer	4—7	8—12	13—20	21 or more
Librarians (n=3)	2	1	—	—	—
Other Staff (n=1)	—	1	—	—	—

Study Hall or Commons (n=4)

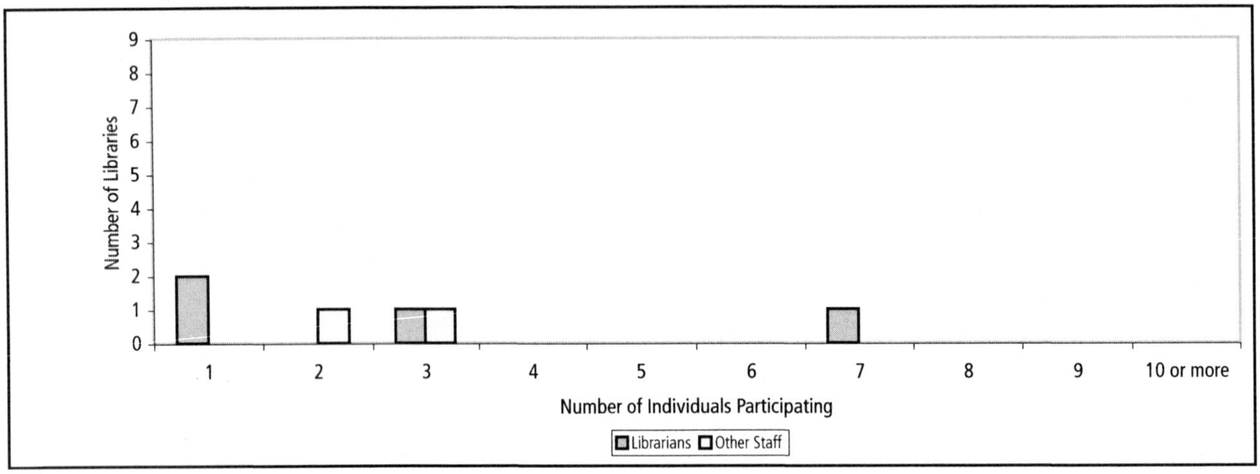

Participating	1	2	3	4	5	6	7	8	9	10 or more
Librarians (n=4)	2	—	1	—	—	—	1	—	—	—
Other Staff (n=2)	—	1	1	—	—	—	—	—	—	—

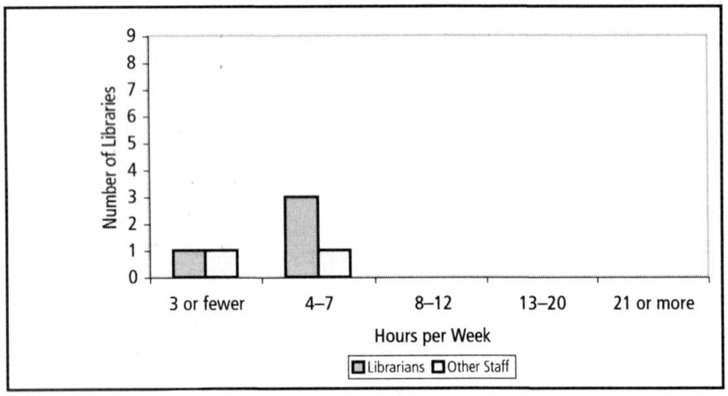

Hours	3 or fewer	4—7	8—12	13—20	21 or more
Librarians (n=4)	1	3	—	—	—
Other Staff (n=2)	1	1	—	—	—

Writing Center (n=3)

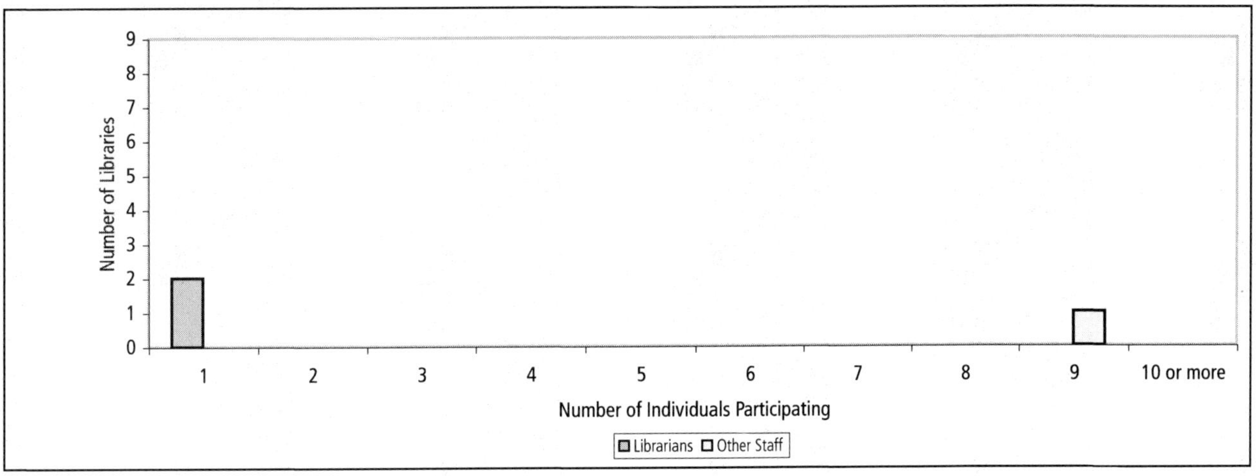

Participating	1	2	3	4	5	6	7	8	9	10 or more
Librarians (n=2)	2	—	—	—	—	—	—	—	—	—
Other Staff (n=1)	—	—	—	—	—	—	—	—	1	—

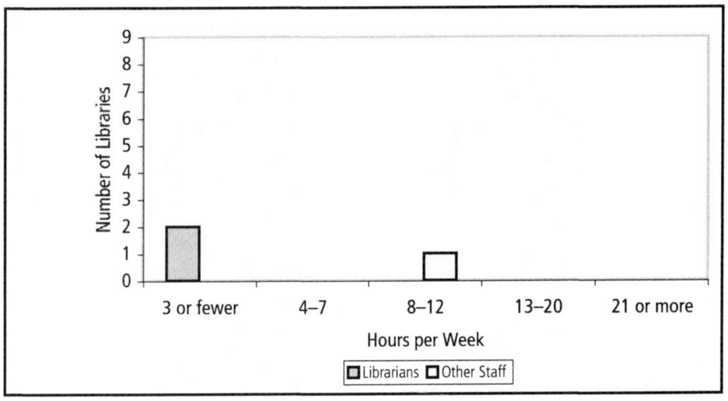

Hours	3 or fewer	4—7	8—12	13—20	21 or more
Librarians (n=2)	2	—	—	—	—
Other Staff (n=1)	—	—	1	—	—

Research Lab (n=2)

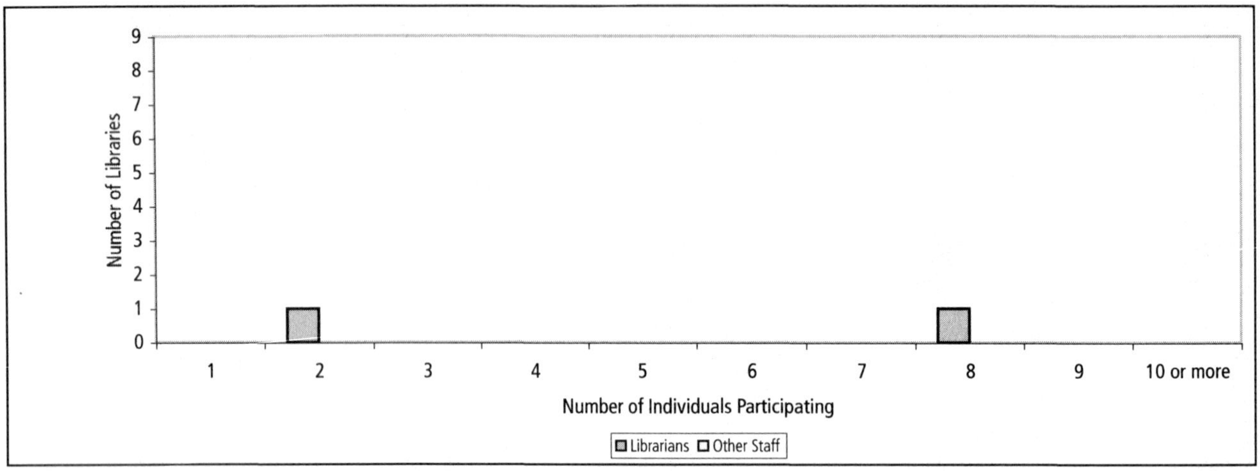

Participating	1	2	3	4	5	6	7	8	9	10 or more
Librarians (n=2)	—	1	—	—	—	—	—	1	—	—
Other Staff (n=0)	—	—	—	—	—	—	—	—	—	—

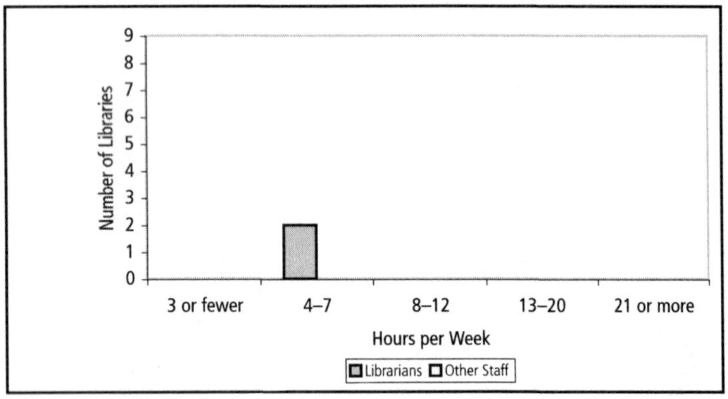

Hours	3 or fewer	4—7	8—12	13—20	21 or more
Librarians (n=2)	—	2	—	—	—
Other Staff (n=0)	—	—	—	—	—

Other Location, please specify (n=7)

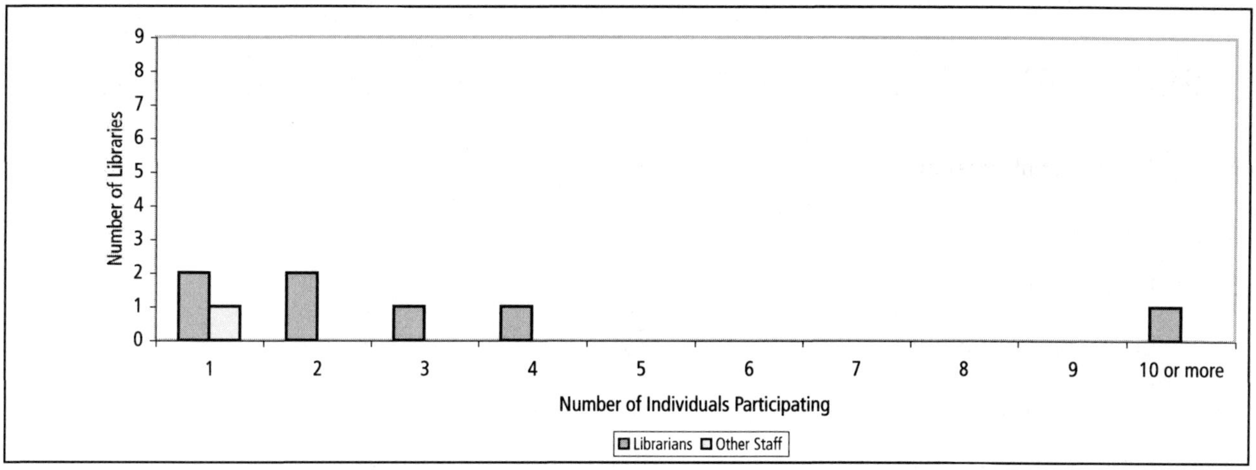

Participating	1	2	3	4	5	6	7	8	9	10 or more
Librarians (n=7)	2	2	1	1	—	—	—	—	—	1
Other Staff (n=1)	1	—	—	—	—	—	—	—	—	—

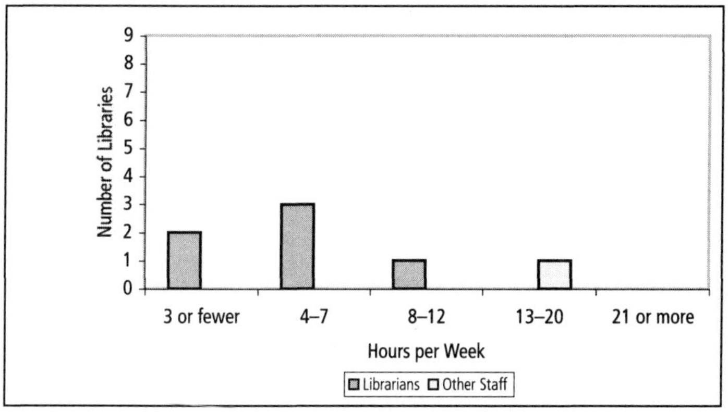

Hours	3 or fewer	4—7	8—12	13—20	21 or more
Librarians (n=6)	2	3	1	—	—
Other Staff (n=1)	—	—	—	1	—

9. In addition to providing services at non-library locations, what other responsibilities do participating library staff members have in the library? Check all that apply. (n=38)

Librarians (n=38)

Reference/information	37	97%
Instruction	37	97%
Departmental liaison	36	95%
Collection development	35	92%
Access services	11	29%
Information technology	9	24%
Administration	8	21%
Technical services	3	8%
Other service, please specify	—	—

Other library staff, including student assistants (n=13)

Reference/information	9	69%
Instruction	8	62%
Access services	2	15%
Information technology	2	15%
Departmental liaison	1	8%
Collection development	1	8%
Technical services	1	8%
Administration	—	—
Other service, please specify	5	38%

Other responsibilities include classroom equipment support, DSpace content coordinator, and document delivery.

10. If your library has a special name or title for the librarians and library staff who provide these services, please enter it below. (n=11)

Information & Education Services Librarian

Field Librarians

Mobile Librarian

Classroom Technology Services staff

Subject Librarians

College Librarians

Clinical Medical Librarians

Liaison Librarian

Outreach Librarian

How are staff members identified to users when they are on duty in these locations (i.e., badges, uniform, sign at a desk, etc.)? (n=27)

Signage	13	48%
ID Badge	6	22%
Known in department	6	22%
Presence at desk/in office	5	19%
Flyers/brochures	3	11%

11. Please indicate when services are offered in each location. Check all that apply. (n=37)

	Weekday daytime	Weekday evening	Saturday	Sunday	Summer session	Cumulative Responses
Office or work space in department	28	2	2	—	6	28
Hospital or other clinical	10	3	1	1	3	10
Computer lab	10	3	1	—	2	10
Residence hall	3	5	1	—	1	8
Study hall or commons	2	2	—	—	—	4
Career center	4	2	1	—	1	5
Student union	3	3	—	—	—	5
Writing center	1	2	1	—	—	3
Research lab	2	1	_	—	1	1
Other location, please specify	8	4	1	—	3	3
Cumulative Responses	35	17	7	1	9	

Other locations:

Weekday daytime	Lecture halls, building lobbies, classrooms, conference rooms, continuing education, department lounges
Weekday evening	student athlete study center, classrooms, continuing education
Saturday	Continuing education
Summer session	Classrooms, conference rooms, continuing education

12. What type of workspace do library staff have in each location? Check all that apply. (n=36)

	Separate office	Shared office	Assigned desk in open area	Unassigned space as available	No assigned space, staff person "roams" or "roves"	Other workspace, please specify	Cumulative Responses
Office or work space in department	15	8	6	5	7	4	27
Hospital or other clinical	2	—	2	3	2	2	10
Computer lab	—	—	1	5	7	—	2
Residence hall	—	—	2	3	1	2	11
Study hall or commons	—	—	2	1	1	—	3
Career center	—	—	2	—	3	1	5
Student union	—	—	—	2	2	1	8
Writing center	—	1	1	—	1	—	4
Research lab	—	—	—	2	—	—	4
Other location, please specify	—	—	2	1	1	1	6
Cumulative Responses	15	9	12	11	14	9	

Other workspace includes:

In department	Breakout room for business, conference room booked for the occasion
Hospital or other clinical	Computer lab, instruction room
Residence hall	Dorm lounge
Other locations	Assigned space in student athlete study center Unassigned space in classrooms and conference rooms Roaming in lecture halls and classrooms Computer lab in continuing education

13. What equipment do library staff use in each location? Check all that apply. (n=35)

	Desktop PC	Laptop PC	Printer	Phone	PDA	Other, please specify	Cumulative Responses
Office or work space in department	22	12	15	15	6	4	25
Hospital or other clinical	6	6	3	3	3	2	8
Computer lab	9	6	2	1	1	3	2
Residence hall	2	6	—	—	—	1	12
Study hall or commons	—	4	—	—	—	—	3
Career center	3	3	—	—	—	—	5
Student union	1	2	—	—	1	1	7
Writing center	2	1	1	—	—	—	3
Research lab	1	2	1	1	1	1	4
Other location, please specify	4	4	—	—	1	2	7
Cumulative Responses	28	18	17	16	9	8	

Other equipment includes:

Office or work space in department	Faculty or researcher's own equipment, projector, tablet PC, LCD projector, network printing to academic department as well as to the library
Hospital or other clinical	Faculty or department equipment, projection equipment
Computer lab	Faculty or department equipment, projection equipment, document camera, overhead projector, LCD projector
Residence hall	Occasionally a projector
Student union	Projector
Research lab	Faculty or department equipment, projection equipment
Other locations	Desktop PC in student athlete study center, classrooms, conference rooms, and continuing education Laptop PC in lecture hall, building lobbies, classrooms, conference rooms PDA and projector in lecture hall

FUNDING

14. How are services in each of these locations funded? Check all that apply. (n=37)

	Library budget	Department/ Agency budget	Other institutional budget	Student fee	Grant or Development Funds	Other, please specify	Cumulative Responses
Office or work space in department	27	4	—	—	—	2	28
Hospital or other clinical	8	1	—	—	—	—	8
Computer lab	12	—	—	—	—	—	12
Residence hall	8	—	—	—	—	1	8
Study hall or commons	3	—	—	—	—	—	3
Career center	3	—	—	—	—	—	3
Student union	4	—	—	—	—	—	4
Writing center	3	—	1	—	—	—	3
Research lab	2	—	—	—	—	—	2
Other location, please specify	7	1	1	1	—	1	7
Cumulative Responses	37	6	2	1	—	4	

Other funding includes:

Office or work space in department	Departmental space is provided free to the library; library budget pays only for salary. Treated as a reference desk shift—equipment is available from the department.
Residence hall	Dorm pays for pizza.
Other location	Library budget supports services in lecture hall, student athlete study center, classrooms, conference rooms, department lounges, faculty offices. School of Continuing and Professional Studies supports services in continuing education. Department budgets, other institutional budgets, and student fees support services in classrooms.

Did the library receive new funding to provide this service?

Yes	3	8%
No	36	92%

If yes, please explain.

Selected Comments from Respondents

"Partial funding for the Field Librarian Program. No funding for other programs."

"A wide range of expanded classroom services have been made possible by new funding from the student educational technology fee, classroom improvement funds, and capital project funds."

"E-rate, donations."

If services are funded by non-library sources (either totally or partially), what is that funding used for? Check all that apply. (n=8)

Equipment	6	75%
Librarian time	4	50%
Advertising	3	38%
Other library staff time	2	25%
Other, please specify	2	25%

Non-library funding also covers Web development, brochures, flyers, and travel. At one institution the colleges provide office space and in some cases have provided equipment.

SERVICE PROMOTION

15. Please indicate how services in each of these locations have been promoted or marketed. Check all that apply. (n=34)

	E-mail	Flyers or posters	Library orient and instruct	Library Web page	Department Web page	Library newsletter	Newspaper	Other, please specify	Cumulative Responses
Office or work space in department	24	14	17	11	11	7	2	6	26
Hospital or other clinical	10	6	6	7	7	4	2	1	11
Computer lab	8	6	7	5	2	2	1	3	10
Residence hall	2	5	3	2	1	—	1	2	8
Study hall or commons	1	2	2	3	1	—	1	—	3
Career center	2	2	2	2	—	—	—	—	3
Student union	2	3	2	3	1	1	3	1	5
Writing center	3	2	2	1	1	1	—	—	4
Research lab	2	1	2	2	1	1	1	1	2
Other location, please specify:	5	2	4	6	2	2	1	1	6
Lecture hall	1	1	1	1	—	1	1	—	1
Faculty offices	1	—	1	2	—	—	—	—	2
Business school	1	—	—	1	—	—	—	—	1
Classrooms	1	—	1	1	1	1	—	—	1
Department lounges	1	—	—	—	—	—	—	—	1
Continuing education	1	1	1	1	1	—	—	1	1
Cumulative Responses	31	24	23	23	16	11	5	8	

Other promotion includes:

Office or work space in department	Door to door, phone calls, direct outreach and departmental newsletter, departmental meeting announcements, sign posted
Hospital or other clinical	Faculty newsletter

Computer lab	Phone calls, direct outreach, personal Web page
Residence hall	Word of mouth, through Academic Programmers
Student union	Center for Adult Learners Conference publicity
Research lab	Phone calls
Continuing education	Networking

SERVICE EVALUATION

16. Please indicate which methods have been used to evaluate the effectiveness of the services in non-library locations. Check all that apply. (n=35)

Informal feedback	25	71%
Observation	16	46%
Surveys	12	34%
Interviews	6	17%
Focus groups	3	9%
Have not evaluated so far	7	20%
Other, please describe	4	11%

Other evaluation methods include statistics of use and evaluation form (2 responses each).

17. List up to three benefits of offering services in non-library spaces. (n=35)

Increased Visibility for Library Staff and the Libraries

"Makes librarians more visible, extends library services to folks who might not be aware of them."

"More visibility for the library staff."

"More prominent role of librarians, able to see user's problems/challenges, increases visibility of library services."

"More face-to-face contact with students."

"Increased visibility for librarian/library."

"Increased visibility of the libraries for all users in the Faculty of Engineering."

"Marketing library services and reaching new (freshman) students."

User Convenience

"Just-in-time service convenient for users. Opportunity to provide instruction on using resources while finding needed information."

"Lets the user use her/his own personal computer where they can take notes. Saves their time and effort."

Put the librarians where the students are, rather than waiting for the students to come to us."

"Convenience for user, saves patrons time."

"Serves audience in their work setting."

"Customized instruction to specific assignments or professional needs. More time to work individually with users."

"Work with users at the point of need."

Relationship Building

"Collegial relationship with faculty, accessibility to faculty."

"Spontaneous conversation with faculty, students, education opportunities. Connecting with faculty and students who don't get to libraries."

"Builds relationships, develops trust. Personal/individualized service is available."

"Fosters 'goodwill' between library and academic department."

"Establishing relationships with faculty and students, including learning more about the departments."

"Easier to interact with faculty & particularly graduate students. Have a better understanding of the inner workings of colleges and departments. Can better influence decisions about curriculum matters by knowing faculty and promoting oneself to serve on college committees."

"Increased collaboration with faculty on projects. Increased awareness of collection needs. Greater awareness of library services and collection services."

"Builds stronger relationships with faculty and students."

"Stronger ties between business school and library; the library is seen as part of business school curriculum."

"Ability to 'connect' at point of need with students in the same physical space in which they are attending class. Ability to 'connect' with faculty in spontaneous conversations. Ability to familiarize ourselves with the unique technical environment and resources of [the department]."

"Ongoing development and strengthening of liaison relationship with the faculty."

Other Benefits

"Serendipity. Information about the technical resources and intellectual environment in the department can help the librarian enhance services (seeing, for example, that everyone uses a Mac.) The nature of library reference work is reinforced on their turf. Responds to a need and could lead to any potential valuable interaction."

"Increased awareness and use of online engineering resources. Increased awareness of services provided by the Science & Technology Library, and the engineering librarians. Savings of time and money through increased access to and use of information resources purchased by the library. Strengthening awareness among library staff of the changing needs of the Faculty of Engineering. Increased use of the libraries' services, particularly the online resources. More one-on-one and small group library and research skills instruction."

"Better integration of library services into academic departments. Reaching people, especially graduate students and faculty who don't visit libraries as often as they used to. For hospital services, identifying information needs at point of care."

"New students who have never used library building are amazed to see what library has to offer. Faculty who are too busy to come to library. Librarians are excited to see how eager people are to contact them."

"Introduce users to new and existing resources in their spaces."

18. List up to three challenges of offering services in non-library spaces. (n=34)

Space

"Getting space. Getting space on a regular schedule."

"Requires more preparation and organization if space is not permanent."

"Needs a lot more time to contact department about equipment, setting up tables, etc."

"Getting space; negotiating for space and equipment."

"Flexibility of space."

"Finding appropriate space in departments has been a challenge."

Staffing/Scalability/Sustainability

"Finding the 'right' individual who can handle the split position."

"Having sufficient staff and time to offer the services—often need to redesign other (i.e., reference desk) service hours."

"Workload/workflow management. Identifying best times to offer service."

"Scheduling of time and balancing demands for 'clerical'/routine activities (never done in department!) with intangible value of building faculty relationships."

"Another added responsibility."

"Scheduling."

"Hard to anticipate how busy it will be. Difficult to determine the best hours."

"Scheduling time in non-library spaces (coordinating schedules, staffing.)"

"Not able to offer consistent services across all schools and colleges; some are open to it and some are not. Arranging logistics can be challenging. Staffing and ramping up to begin a true program."

"For large colleges, providing adequate number of librarians with some subject expertise. Balancing the time and commitments for colleges-based duties and parent organization."

"Pulls those librarians out of the rotation for staffing in-house reference desks. Keeping the services in non-library spaces a priority in busy times. Scalability—how can we do this for every department?"

"Sustainability over time, competing demands on librarians' time."

"Getting enough transactions to make it worthwhile, and traveling around campus."

"Balance with scheduling of staff at traditional service points."

"Funding—serves a relatively small population and is very labor intensive. Need to quantify the value of the service and do cost/benefit analysis to see if service is sustainable."

Marketing

"Need to advertise in many locations and through variety of venues, formats. May not be readily associated with the library."

"By late 2004, all engineering departments are to be located at western edge of campus. We anticipated a reduction in engineering clientele in the local campus SciTech library. We wanted to maintain and build upon excellent working relationship between SciTech library and Faculty of Engineering. We wanted to keep the library's profile strong among engineering clientele."

"Promoting the service, especially getting onto department's Web site."

"Introducing the concept and making the user aware of the possible service."

"Continuous promotion."

"Promoting service to users."

Integrating Important Print Resources Into Service

"Being away from library print resources."

Being away from the print collection."

"Harder to show and use important print resources."

"Difficult to show more than basic services in public spaces."

Other

"Learning the cultures of the departments."

"Technological issues (IT support if needed unexpectedly). More demands for special service."

FUTURE PLANS

19. Does your library anticipate offering scheduled in-person library services in non-library spaces within the next three years? Example of such services included scheduled office hours, reference, consultation, workshops, demonstrations, technical assistance, and clinical support. Do not include course instruction or other ad hoc activities.

	Haven't offered services outside the library		Currently offer services outside the library	
Yes	6	18%	35	90%
No	28	82%	4	10%

If yes, please indicate in which locations library services might be offered. Check all that apply.

Haven't Offered Services Outside the Library (n=6)

	Reference	Individual Consultation	Group Consultation	Workshops/ Demos	Technical Assistance	Clinical Support	Other Service, please specify	Cumulative Responses
Office or work space in department	4	4	3	3	4	—	—	4
Study hall or commons	3	3	3	2	1	—	—	3
Residence hall	3	2	2	2	1	—	—	3
Student union	3	2	2	1	—	—	—	3
Computer lab	—	1	—	1	1	—	—	2
Writing center	—	—	—	1	1	—	—	1
Career center	—	1	1	1	1	—	—	1
Research lab	1	1	1	—	1	—	—	1
Hospital or other clinical	—	—	—	—	—	—	—	—
Other location, please specify	1	1	1	—	1	—	—	1
Cumulative Responses	5	5	4	4	4	—	—	

The other location for future services is the atrium of a building complex.

Currently Offer Services Outside the Library (n=33)

	Individual Consultation	Reference	Workshops/ Demos	Group Consultation	Technical Assistance	Clinical Support	Other Service, please specify	Cumulative Responses
Office or work space in department	23	17	19	18	12	2	—	23
Residence hall	14	11	11	11	6	—	—	17
Computer lab	6	6	9	7	5	1	—	11
Student union	6	8	5	3	3	—	—	10
Writing center	6	5	7	6	3	—	1	9
Study hall or commons	7	6	5	6	3	—	—	7
Career center	3	4	6	2	1	—	—	7
Hospital or other clinical	6	5	6	6	4	4	—	6
Research lab	3	2	3	3	4	—	—	5
Other location, please specify	4	3	3	1	2	—	—	5
Lecture hall	—	—	1	—	—	—	—	1
Student athlete study center	1	1	—	—	—	—	—	1
Classrooms	1	—	1	1	—	—	—	1
Continuing education	1	1	1	—	1	—	—	1
Department lounges	1	1	1	—	1	—	—	1
Cumulative Responses	32	27	26	24	17	5	1	

ADDITIONAL COMMENTS

20. Please submit any additional information regarding library services in non-library spaces at your institution that may assist the authors in accurately analyzing the results of this survey. (n=26)

Selected Comments from Respondents

"The library is planning to collaborate with the Office of Information Technology to cross-train computer lab proctors. The plan is to have the lab proctors spend a semester at a subject reference desk in the library, science for instance, prior to proctoring a lab in the science building. These labs, which are located in various locations around campus would then be designated as Electronic Branch Libraries."

"There are many library series that take place outside the library that are occasional, not ongoing, e.g., connecting with frats/sors in their houses, talking to a group of students from a particular residence. We are constantly looking for ways to connect outside the library walls."

"Our portable 'Information Station' will pilot this summer in four campus locations. It consists of a table with colorful canopy and a wireless computer."

"Offering some formalized library service in a 'Manoa Village' residential dorm setting is in the brainstorming stage."

"Business librarian's comments: I have many faculty contacts during office hours. There's a high proportion of service to M.B.A. and Ph.D. students. The use of a tablet PC identifies librarians as users of advanced technologies. Previous education librarian's comments: This service allowed students to get assistance between classes. Many of the education grad students came in for late afternoon classes and didn't have time to get to the library. By setting up shop in their area they could get assistance before going to their class. Each session there were drop-in faculty members who normally wouldn't have gone to the library. This allowed for much broader liaison and research assistance to faculty members. We would both highly recommend office hours for each subject liaison."

"We presently offer many of the above services on an ad hoc basis; i.e., we don't have regularly scheduled hours but respond instead to user demand and/or need. The level of assistance we are able to provide in the future on a regularly scheduled basis will be dependent on the funding received for full- and part-time staff."

"We have no plans for reducing the number of satellite libraries, (presently 8 of 15) so there's good proximity of library service and communities in many of the social, life, and physical sciences. We have engaged schools without embedded libraries about housing outposts for reference-related services. The schools perceived the need and benefit, but space, equipment, and networking constraints have always been impediments. These conditions are part of a campus-wide problem. One effort to address that problem has involved the expansion of teaching and consultative spaces within the large libraries. The main library has undergone extensive renovation, as has the fine arts library. Future upgrades are planned for biomedicine, another facility that serves multiple communities. We're also using the libraries as the nexus for a more formal integration of information and academic support services. While this effort unfolds, we are finding a general increase in the number and variety of ad hoc service events (consultations and the like) taking place in faculty offices and residential facilities. Until this year we did train and supervise a large corps of student library advisers who lived and worked in the major colleges houses. That effort has been revamped using chat and other forms of outreach."

"Our current model has 'librarians with laptops' in offices and labs on a by-appointment basis. It is a 'just in time' rather than a 'just in case' model."

"These activities have been trial balloons and fairly ad hoc thus far, however, organizing library activities in non-library spaces is called for in our strategic plan and is likely to be a priority as we move forward."

"The university is planning a new program for residential colleges which will begin in 2008. The first year of the program is to be establishment of a Freshman Commons, and the library is involved with the planning for that. We do anticipate providing library services in non-library spaces within the context of that developing program."

"Health science library wants to get affiliate appointments for librarians engaged in the service. Main library wants to grow 'office hours' model to establish a beachhead for expansion of service—more hours, more participants, more time during week spent in departments."

"This is difficult for us to answer as a public library. I would say that in general we are trying to reach out to our users, through off-site service delivery, electronic resources, etc."

RESPONDING INSTITUTIONS

University of Alabama
University at Albany, SUNY
University of Alberta
University of Arizona
Arizona State University
Auburn University
Boston College
Boston Public Library
Boston University
Brigham Young University
University at Buffalo, SUNY
University of California, Davis
University of California, Irvine
University of California, Los Angeles
University of California, Riverside
University of California, San Diego
University of California, Santa Barbara
University of Chicago
Colorado State University
Cornell University
Dartmouth College
Emory University
University of Florida
Florida State University
George Washington University
University of Guelph
University of Hawaii at Manoa
University of Houston
Indiana University Bloomington
University of Iowa
Iowa State University
Kent State University
University of Kentucky
Library and Archives Canada
Library of Congress
Louisiana State University
McGill University
McMaster University
University of Manitoba
University of Maryland
University of Massachusetts, Amherst
University of Miami
University of Michigan
Michigan State University
University of Minnesota
University of Nebraska–Lincoln
New York Public Library
University of North Carolina at Chapel Hill
North Carolina State University
Northwestern University
University of Oklahoma
Oklahoma State University
University of Oregon
University of Pennsylvania
Pennsylvania State University
University of Pittsburgh
Purdue University
Rice University
University of Rochester
Rutgers University
Smithsonian Institution
Southern Illinois University Carbondale
Syracuse University
University of Tennessee
University of Texas at Austin
Vanderbilt University
University of Virginia
Virginia Tech
University of Washington
Washington State University
University of Waterloo
Wayne State University
University of Western Ontario
University of Wisconsin–Madison
Yale University

REPRESENTATIVE DOCUMENTS

Publicizing the Service

Subject: Brett's Office Hours
Date: Tue, 14 Oct 2003 12:02:16 -0500
From: Brett Spencer <dbspence@bama.ua.edu>
To: rsigler <rsigler@cj.as.ua.edu>, rspano <rspano@bama.ua.edu>, jwilliam <jwilliam@bama.ua.edu>, ijohnson <ijohnson@bama.ua.edu>, clo <clo@bama.ua.edu>, rfarmer <rfarmer@bama.ua.edu>

Dear Criminal Justice Faculty:

I am pleased to announce a new library service for the Department of Criminal Justice. As many of you may know, I am keeping office hours in Farrah Hall (Office 417) this semester at the following times:

Mondays 10:00--12:45
Thursdays 12:00--1:00

I am doing this to make it more convenient for CJ students to obtain in-depth research help: I scheduled the office hours when a number of classes meet in Farrah Hall so that students in those classes can drop by Office 417 before and after their classes. I have a computer in this office that the Library lent me and can thus provide a great deal of research assistance by searching online databases, catalogs, legal documents, and journals. (Earlier this semester I offered office hours in the graduate student lab, but now I have my own office so that I can offer this service to undergraduates as well).

Thus, please feel free to refer undergraduate and graduate students to Office 417 during these hours, and I will be glad to assist them with finding sources for their papers. Of course, students can still schedule appointments at other times at the Gorgas Library by contacting me at 348-1200 or dbspence@bama.ua.edu.

Regards,
Brett Spencer
Criminal Justice Librarian
Box 870266
Gorgas Library--Information Services
University of Alabama
Tuscaloosa, AL 35487-0266
(205) 348-1200

Subject: Library Information
Date: Thu, 22 Jan 2004 14:21:25 -0600
From: Brett Spencer <dbspence@bama.ua.edu>
To: rsigler <rsigler@cj.as.ua.edu>, rspano <rspano@bama.ua.edu>, jwilliam <jwilliam@bama.ua.edu>, ijohnson <ijohnson@bama.ua.edu>, clo <clo@bama.ua.edu>, rfarmer <rfarmer@bama.ua.edu>, klemmac@bama.ua.edu, lharris@shletonstate.edu, pauljo.mills@att.net, lolin@bama.ua.edu, DarindaPoe@hotmail.com, globetti1@earthlink.net, phallida@bama.ua.edu, Charles024@yahoo.com, DPeck@tenHoor.as.ua.edu

Dear CJ and Sociology faculty:

I hope that you are having a good spring. I wanted to email you with some library information for this semester. In the way of new resources, the Library has acquired the following databases that are now available on the databases page http://www.lib.ua.edu/resources/databases/ :

Communication and Mass Media Complete--provides the most robust, quality research solution in areas related to communication and mass media. Access this by clicking "Ebscohost" on the databases page, scrolling down, checkmarking this database, and clicking "Continue" at the bottom of the page.

Military and Government Collection--Designed to offer current news pertaining to all branches of the military and government, this database offers a thorough collection of periodicals, academic journals, and other content pertinent to the increasing needs of those sites. Offers some policing and political science journals.

Proquest Newspapers--contains the following:
Atlanta Journal and Constitution, January 1995-
Christian Science Monitor, October 1995-
New York Times, March 1998-
New York Times Magazine, April 1993-
Wall Street Journal, October 1997-
Washington Post, February 1996-.

Please let me know if there are any books or videos that you would like the library to order for your classes or research.

It is also my pleasure to continue to serve you and your students in Farrah Hall as I did last semester. This semester I will be offering office hours in Office 417 in Farrah Hall during the following times:

Mondays 9-11
Tuesdays 12-1

This research service in Farrah is for both graduate and undergraduate students. Students can also contact me at 348-1200 or dbspence@bama.ua.edu for appointments at the Gorgas Library.

Please let me know if I can be of assistance with anything.

Thank You Very Much,
Brett Spencer
Reference Librarian; Liaison to the Department of Criminal Justice
Box 870266
Gorgas Library--Information Services
University of Alabama
Tuscaloosa, AL 35487-0266
(205) 348-1200

http://www.library.ualberta.ca/subject/engineering/onsite/index.cfm

Home | Catalogue | Databases | Ejournals | Subject | Services | About Us | My Account

Home > Subject > Engineering > Librarian On-Site!

Version française

Librarian On-Site
An Information and Reference Service for Engineering Students and Staff

A collaborative effort between the *University of Alberta Libraries* and the *Faculty of Engineering*, **Librarian On-Site** brings the information and reference services of the Cameron Science and Technology Library to all students, faculty, researchers and staff in the Faculty of Engineering. The service ran as a pilot project from September 24th, 2003 until April 7th, 2004 in the Engineering Teaching and Learning Complex (ETLC), and will begin again this fall.

Engineering librarians **Jerry Kowalyk** and **Randy Reichardt** will be available for consultation:

Location: *E2-006 ETLC*
Days: *Wednesday and Thursday afternoons*
Time: *1:00-3:00 pm (1300-1500 hrs)*
Dates: *September 29th, 2004 - December 1st, 2004*

Background: The field of engineering has been a leader in the conversion of research materials to electronic format. The University of Alberta Libraries makes available many of the major electronic reference materials in this field, with more on the way. With the impending completion of the new engineering buildings, all departments within the Faculty of Engineering will have moved to the western edge of the campus by late 2004, and away from the Cameron Science and Technology Library. As a result, we anticipate seeing a reduced number of Engineering clientele in the library.

The University of Alberta Libraries recognizes the importance of developing strong library and research skills, and will provide an opportunity for students working in the ETLC to consult with librarians about finding information resources to support their projects and assignments. At present, library and research skills instruction is offered in at least 25 different courses in the departments and schools within the faculty. This new initiative will enhance the strong relationship shared between the Faculty of Engineering and its Departments, and the U of A Libraries, and help maximize the use of the Libraries' resources.

For more information, please contact:

Randy Reichardt, Public Services Librarian (Engineering)
randy.reichardt@ualberta.ca / 492-7911

Last updated: 29 July 2004

Ask Us A Question
» Chat » Phone
» Email » In-Person

Print
Email

Ask Us | Contact Us | Site Map | A - Z Index | Site Search | RSS Libraries is a division of Learning Services

UNIVERSITY OF ALBERTA Copyright © 2004 | University of Alberta | Privacy Policy

Librarian On-Site!

Would you like to learn more about library and online resources in your subject area?

Do you need help finding information for your:

- *Design project?*
- *Thesis research?*
- *Research paper?*
- *Library assignment?*
- *Project proposal?*

Please drop by and see us, we'll be happy to help!

— Librarian On-Site Assistance —

ETLC Computer Laboratory E2-006

Pilot project: Beginning Sept 24th, to Dec 2003

Every Wednesday and Thursday, 1:00-3:00 pm

Brought to you by the Librarians of the Cameron Science & Technology Library

University of Alberta
LIBRARIES

ENGINEERING
U of A

Letter to editor of student magazine to advertise UGL.Union

DIALOGUE

Date Rape

Editor:

I am writing in response to the date rape article printed in the October 22 issue of *Generation* ("One Woman's Story). I am so tired of hearing women scream how they were date raped and taken advantage of.

This is what I find amazing: this woman accepted a date and went out to dinner with a man she felt uncomfortable with. Then, invited him to her apartment, drank with him, smoked pot, and let him feel on her, *and* she had no objections when he took her into the bedroom? This girl still has the nerve to plead rape. Where was your voice when it really counted? You should take a good look in the mirror, maybe it's yourself you shouldn't trust, not men.

These females literally make me sick. If she thinks that what that man has done is rape, then she needs to wake up! You only are what you portray yourself to be. Any women who allows a man she has no intention of sleeping with, into her bedroom, to get drunk and high, must not think too highly of herself. And guess what? Neither does he! Don't blame the man for having sex with you when it was practically offered to him.

Anyone who takes such irresponsible actions suffers the consequences. Please don't come to us for sympathy. Sorry if reality hurts. And please don't try to make us believe you want to teach us something. I would like to believe most mature women would know better anyway. It's time to grow up!

Yvonne Kamieniecka
Undergraduate Student

Writing Contest Announced

Editor:

Annually, UB awards a cash prize to a senior student based on submitted creative winning entries. The prize fund was established by the family and friends of Gregory Capasso, a UB student who died in the terrorist bombing of Pan Am Flight 102 over Scotland. The fund also supports an annual Media Studies academic scholarship. Winners are acknowledged on a plaque located in the Student Union.

Interested students can send their entries to Student Affairs, 542 Capen Hall, North Campus. The deadline for entries is December 1, 1996.

Dennis R. Black
Dean of Students

More Than Weird Scarves & Cookies

Editor:

This is a year to celebrate Girl Scouting! Girl Scouts of the USA is celebrating its 85th anniversary in 1997 and the Girl Scout council of Buffalo & Erie County, Inc. is marking 80 years of Girl Scouting in Buffalo.

To commemorate these occasions, the Girl Scout Council of Buffalo & Erie County is searching for women and men living in the area who have ever been a Girl Scout.

Many special events are planned for the upcoming year, including a reunion of local alumni to be held on March 15, 1997. These events will be an opportunity to network, build new contacts and rekindle old acquaintances with Girl Scouts from all walks of life.

If you have ever been a registered Girl Scout, please call Girl Scout Council at 837-6400, ext. 232.

Amy Hammond
Public Information Assistant

UGL Satellite of Love

Editor:

Once again, the Undergraduate Library (UGL) is offering ugl.union, a satellite reference service located on the second floor of the Student Union.

UGL librarians will be available 11AM-1PM every Monday, Tuesday, and Wednesday through November 22 to assist students with research questions, the UB Libraries Catalog, BISON II, and the Internet. Please stop by to see what we have to offer.

If you have any questions about ugl.union please contact Lara Bushallow-Wilbur at LBW@acsu.buffalo.edu or 645-2943 ext. 236.

See you at the Union!

Lara Bushallow-Wilbur

Generation welcomes letters from the University community. Letters must be signed and contain a phone number and University affiliation. Please be brief and to the point. Send letters to Dialogue, c/o *Generation*, Suite 114 Student Union, Buffalo, NY 14260, or to generation@ubvms.cc.buffalo.edu. Campus mail is free. All questions about this section can be directed to the Editor-in-Chief.

When you split a Uranium atom, you get nuclear energy.

When you split an issue of Generation, you just get a torn magazine.

When you're splitting hairs over something, write a letter to us.

http://web.uflib.ufl.edu/ps/librarynews/LibNews0804.pdf

New library catalog incorporates "patron empowerment" features

A new Online Public Access Catalog (OPAC) was unveiled over the summer by the University of Florida Libraries, and the change means more to library patrons than a different look. The system offers new functionality, with added benefits for both the novice and experienced user. Among the new capabilities of the system are "patron empowerment" features, which allow library patrons to sign in to the system and do a number of things they were unable to do in the old LUIS or WebLUIS systems. These new features include accessing your personal account information to view a list of items checked out or overdue items, renewing library materials directly from the catalog, displaying information regarding the status of recalls and hold requests submitted, changing the account password, viewing proxy (formerly deputy) borrower activity, and customizing how search results appear when signed in.

In addition, there are several features that will be very beneficial to more advanced searchers, such as faculty and graduate students, who often work on long-term research projects. When signed in to a personal account, full bibliographic records chosen from the results list can be saved to an "E-Shelf" and persistent search queries that the user may want to re-run later can be saved to a "Saved Searches" location. Both the records and searches are saved permanently on a server and not merely saved in the local location, so they are available indefinitely until deleted by the account holder.

For more information about the new library catalog or other library resources or services, contact the library subject specialist for your area (see list at **http://www.uflib.ufl.edu/selects.html**) or a librarian at any of the library locations on campus.

Patrick Reakes
Journalism & Communications Library

Librarians take reference on the road

The UF Libraries launched a new service during the Summer B semester called the "Information Station." The Information Station is a mobile reference and information booth equipped with a wireless tablet PC that is moving around campus to provide students, faculty, and staff with answers to their questions. During the pilot phase the station was located in three major locations, Turlington Plaza, Plaza of the Americas, and Little Hall/Carlton Auditorium area. The station was staffed by several librarians from the Humanities and Social Sciences Reference department. A variety of questions were asked, ranging from where a building was located to questions about the Library West construction and the retrieval of books from its collection. The Information Station will continue to move around campus during the fall semester helping students adjust to the changes the library has undergone, such as the new library home page and catalog. Watch for the Information Station in your area of campus this fall.

Marina Salcedo
Humanities & Social Sciences Reference

Library News ▶ *Page 3*

UNIVERSITY OF MINNESOTA

http://education.umn.edu/library/default.html

Return to: U of M Home

One Stop | Directories | Search U of M

UNIVERSITY OF MINNESOTA

COLLEGE OF EDUCATION & HUMAN DEVELOPMENT

MOBILE LIBRARIAN *for* EDUCATION

What's inside

University > College > **Mobile librarian**

Schedule appointment

Ask us!

Selected library resources & services

Library workshops

College resources

Academic departments

Research and expertise

Student services

University resources

About the U of M

University libraries

edlib@umn.edu
Mobile Librarian for Education
University Libraries

Search the college

go

UNIVERSITYLIBRARIES

Mobile librarian for education

A mobile librarian is available to meet with individual(s) for library research consultation.

Come to us or ask us to come to you! Mobile librarians are available on a limited basis one afternoon a week for a "house call" in your office, Wilson Library (West Bank), Magrath Library (St. Paul Campus), or other college location.

This service is available to all college staff, faculty, and students.

Schedule an appointment

It all starts here when you sign up for a time to meet. Appointment times are available on a limited basis one afternoon a week.

Other library services

Selected library resources & services

Link to the Libraries' list of recommended sources for information to use in education and human development research.

Ask us!

No time for an appointment? No problem! Submit your question via e-mail and receive a reply within 24 hours. Or chat with a librarian using our live, real-time chat reference service.

Sign-up for a Library workshop

Free and open workshops help students, faculty, and staff with the various challenges they face doing research at the University.

Contact college Webmaster | Contact U of M | Privacy

The University of Minnesota is an equal opportunity educator and employer.

Last modified on November 04, 2004

http://www.lib.utexas.edu/about/news/onlocation.html

News

GENERAL LIBRARIES
The University of Texas at Austin

www.lib.utexas.edu P.O. Box P, Austin, TX 78713 512 495-4350

Librarians take information on the road

Students can now get a burger, fries and information at the Texas Union

AUSTIN, Tx (September 8, 2003) – "Information and information on how to access that information is what they're hungry for," says Damon Jaggars, Head, Undergraduate Library and Deputy Associate Director for Public Services for the General Libraries, as he stands in the lobby of the Texas Union with a wireless laptop on the nearby desk. He emphasizes that the new "Librarians on Location" program is designed to help meet the research and information needs of students in an innovative, timesaving and convenient way.

During this fall semester, librarians with their trusty wireless laptops will move beyond the walls of various General Libraries units and into the Texas Union. Each Tuesday from noon until 4:00 p.m. librarians will be available in the Texas Union lobby next to the Information Center (second level/ground floor) to help students with their research.

Michele Ostrow, Head of General Libraries Library Instruction and Information Literacy Services, and co-coordinator of the "Librarian on Location" pilot project, points out that the program is designed to encourage students to keep the library and its information specialists at the forefront of their research. Ostrow emphasizes that plans are in the works to expand the program into other locations as soon as staffing and schedules permit.

For more information on the program and the locations and times that librarians will be available, consult the Librarian on Location web site [http://www.lib.utexas.edu/services/reference/libonloc.html] or contact Michele Ostrow, Flawn Academic Center (FAC) 250, micheleo@mail.utexas.edu, (512) 495-4534.

# # #

http://www.lib.vt.edu/help/instruct/clprg.html

University libraries

Home | Hours | Ask a librarian

Search Addison, the library catalog: Keyword GO

Find... Services... Help...

Off Campus Sign in

College librarian program

Since 1994, College Librarians have provided the faculty, staff and students of each of Virginia Tech's colleges with information services tailored to the needs of each constituent group. These librarians serve their colleges by such things as:

- Purchasing materials - books, videos, serials, databases - in support of the college;
- Providing subject specific research;
- Helping design realistic assignments that will deepen and test discipline-specific information skills;
- Teaching information skills sessions;
- Leading or arranging tours of the libraries; and
- Assisting with distance education classes.

During 2002-2003, the College Librarians and branch librarians taught over 830 class sessions to over 19,000 students and placed over 25,000 orders for library materials. College Librarians have also served on departmental committees, partnered with faculty in grants and research, assisted graduate degree candidates with literature reviews, and prepared class-specific web pages. For additional information: Instruction and Reference Department.

Bibliography of articles about Virginia Tech's College Librarian Program.

college librarian assignments

College / Subject	Librarian Contact Information
Agriculture and Life Sciences	
Agricultural & Applied Economics • Animal & Poultry Science • Biochemistry • Biological Systems Engineering • Crop & Soil Environmental Sciences • Dairy Science • Entomology • Horticulture	Margaret Merrill 231-9670 • 1330 Litton Reaves 231-2322 • 3340A Newman Library
Human Nutrition, Foods & Exercise • Food Science & Technology • Agricultural & Extension Education • Plant Pathology, Physiology, & Weed Science • Medicine	Caryl Gray 231-9229 • 5005 Newman Library 231-7419 • 117 Wallace
Architecture & Urban Studies	
Architecture • Art & Art History • Building Construction • Industrial Design • Interior Design • Landscape Architecture • Urban Affairs and Planning	Heather Ball 231-9272 • Art & Architecture Library

http://www.lib.vt.edu/help/instruct/clprg.html

Public Administration and Policy	Bruce Pencek 231-2140 • 3033 Newman Library 231-5806 • 311 Lane Hall
Business	
Accounting and Information Science • Business • Finance • Hospitality & Tourism • Insurance & Business Law • Management • Business Information Technology • Marketing	Ellen Krupar 231-4863 • 2061 Pamplin 231-2490 • 3007 Newman Library
Engineering	
Aerospace & Ocean • Chemical • Engineering Science & Mechanics • Materials Science & Engineering • Mechanical	John Cosgriff 231-9248 • 5010 Newman Library
Civil & Environmental • Computer Science • Electrical & Computer • Engineering Education • Industrial & Systems • Mining & Minerals	Larry Thompson 231-8693 • 5010 Newman Library
Liberal Arts & Human Sciences	
English • Language/Linguistics • Literatures in English	Nicole Auer 231-8684 • 3050 Newman Library 231-8446 • 424 Shanks Hall
Communication • Foreign Languages • Theatre Arts • Music • Philosophy • Religion • Women's Studies	Anita Haney 231-4116 • 3030 Newman Library 231-5806 • 311 Lane Hall
Appalachian Studies • Geography • History • Political Science • Science & Technology Studies	Bruce Pencek 231-2140 • 3033 Newman Library 231-5806 • 311 Lane Hall
Educational Leadership & Policy Studies• Human Development • Teaching & Learning	Michelle Young 231-7040 • D4 War Memorial (3rd flr) 231-7419 • 117 Wallace
Apparel, Housing & Resource Management	Caryl Gray 231-9229 • 5005 Newman Library 231-7419 • 117 Wallace
Natural Resources	
Fisheries & Wildlife • Forestry •	Irene Glennon

Geography • Wood Science & Forest Products	231-9250 • 3005 Newman Library 231-6225 • 138D Cheatham
Science	
Biology • Chemistry • Geosciences • Mathematics • Physics and Astronomy • Psychology • Statistics	Ed Lener 231-9249 • 4010 Newman Library 231-2612 • 3066A Derring Hall
Veterinary Medicine	Vicki Kok 231-6610 • Office, Vet Med Library 231-6170 • Vet Med Library

Other contacts

Economics, Law	David Beagle 231-9231 • 3035 Newman Library
Military Science	Lane Rasmussen 231-4645 • 304 Newman Library
Government Documents	
GIS & Cartographic Data • Maps • U.S. Federal • Virginia Information Resources	Bruce Obenhaus 231-6181 • 2005 Newman Library
Sociology, Africana Studies	Paul Metz 231-5663
University Studies	Nan Seamans 231-2708 • 2032 Newman Library

Last updated: 8/13/04 by Robert Sebek

Virginia Tech Blackboard at Virginia Tech MYVT

University Libraries
Virginia Tech
P.O. Box 90001
Blacksburg, VA 24062-9001
540-231-6170

http://www.cob.vt.edu/directory/staff.htm

Virginia Tech

PAMPLIN
COLLEGE OF BUSINESS

Search VT
Go

Staff Directory

OFFICE OF THE DEAN
1030 Pamplin Hall
Blacksburg, VA 24061
(540) 231-6601

Richard E. Sorensen, Dean

- Albimino, Sam, Director of Corporate Relations
- Bonham, T.W., Associate Dean, Administration & Research, (540) 231-6595
- Crowther, O.T., Director of College Development, (540) 231-5662
- Griffith, Barbara T., Accountant, Administration & Research, (540) 231-6595
- Ho, Sookhan, College Communications Manager, (540) 231-5071
- Jones, Denise, Administrative Staff Assistant, Dean's Office, (540) 231-6601
- King, John, Associate Director of Alumni Relations, (540) 231-9551
- Krupar, Ellen, College Librarian for Business, (540) 231-4863
- Moore, Barry N., Director, Management and Professional Development, (540) 231-5567
- Pokorski, Dale M., Director of Computing, (540) 231-7377
- Politis, John, Director, Business Technology Center, (540) 231-2730
- Sarver, Sharon L., Program Support Technician, Management and Professional Development, (540) 231-5566
- Scott, Sharon G., Associate Director, Management and Professional Development, (540) 231-6601
- Koehler, Heidi, Assistant to the Director of Development, (540) 231-6601

ACADEMIC DEPARTMENTS

Accounting and Information Systems - 3007 Pamplin Hall, (540) 231-6591

- Brown, Robert M., Department Head, (540) 231-6591
- Hawthorne, Anna., Secretary, (540) 231-6591
- Neece, Phyllis I., Head Secretary, (540) 231-9639
- Perfater, Arnita S., Secretary, (540) 231-6592

Business Information Technology - 1007 Pamplin Hall, (540) 231-6596

- Taylor, Bernard W., Department Head, (540) 231-6596
- McCoy, Tracy J., Head Secretary,(540) 231-9630
- Long, Teena P., Secretary, (540) 231-6597
- Seavey, Sylvia N., Secretary, (540) 231-6596

Finance - 1016 Pamplin Hall, (540) 231-5904

- Singal, Vijay , Department Head, (540) 231-5700
- Acton, Julia S., Head Secretary, (540) 231-4551
- Blankenship, Dianne M., Secretary, (540) 231-7251
- Brownlee-Bowen, Leanne, Office Services Specialist, (540) 231-5886

Service Schedule Information

THE LIBRARIAN IS IN at FURCOLO

- Do you need help doing **research**?
- Are you interested in using the **Internet** to find education-related materials?
- Would you like assistance navigating the **Library's computerized network**?

Then come to one of the sessions below:

Where: Mac Computer Lab in the basement of **Furcolo Hall**
When: Tuesdays March - May
Time: 9:30 - 11:00 a.m.

No appointment necessary Contact Lori Mestre (Education Reference Librarian) if you prefer to set up an alternate appointment.

lori.mestre@library.umass.edu 545-6851

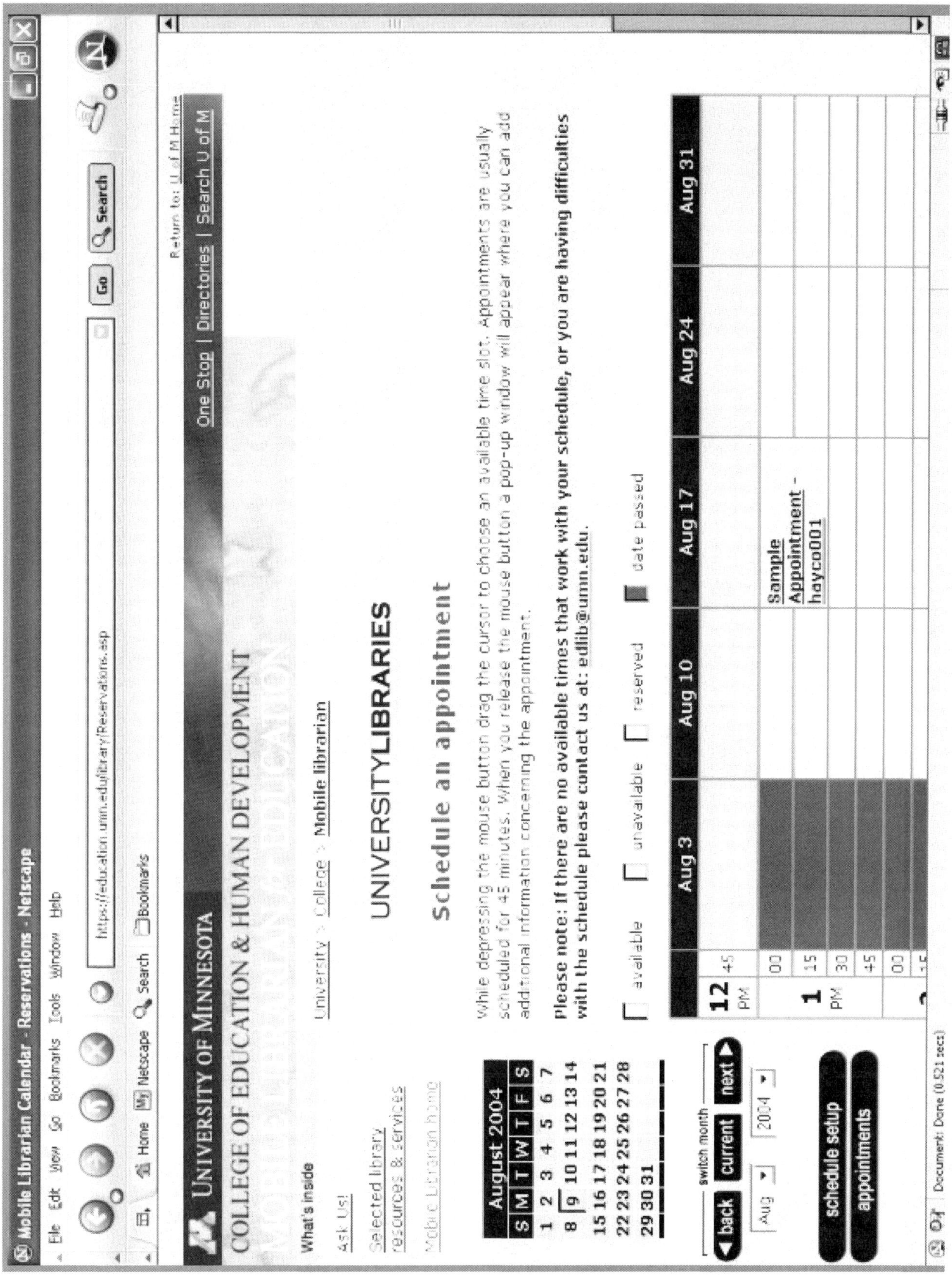

Mobile Librarian Calendar - Reservations - Netscape
File Edit View Go Bookmarks Tools Window Help
https://education.umn.edu/library/Reservations.asp
Go
Search
Home
Netscape
Search
Bookmarks
UNIVERSITY OF MINNESOTA
COLLEGE OF EDUCATION & HUMAN DEVELOPMENT
Return to: U of M Home
One Stop | Directories | Search U of M
What's inside
Ask Us!
Selected library resources & services
Mobile Librarian home
University > College > Mobile librarian
UNIVERSITYLIBRARIES
Schedule an appointment
While depressing the mouse button drag the cursor to choose an available time slot. Appointments are usually scheduled for 45 minutes. When you release the mouse button a pop-up window will appear where you can add additional information concerning the appointment.
Please note: If there are no available times that work with your schedule, or you are having difficulties with the schedule please contact us at: edlib@umn.edu.
available
unavailable
reserved
date passed
August 2004
S M T W T F S
1 2 3 4 5 6 7
8 9 10 11 12 13 14
15 16 17 18 19 20 21
22 23 24 25 26 27 28
29 30 31
switch month
back
current
next
Aug
2004
schedule setup
appointments
Aug 3
Aug 10
Aug 17
Aug 24
Aug 31
12 PM
1 PM
Sample Appointment - hayco001
Document: Done (0.521 secs)

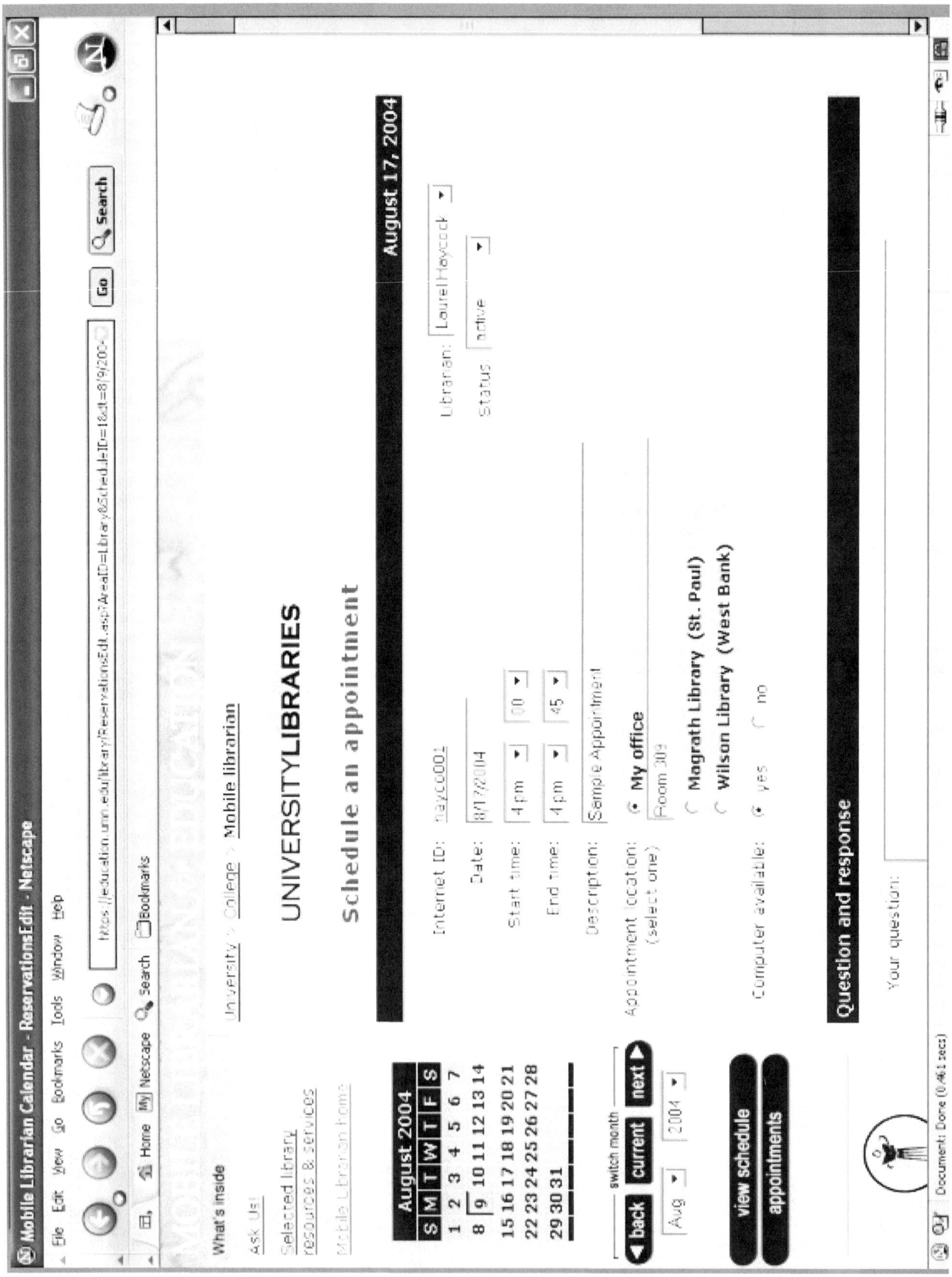
Mobile Librarian Calendar - ReservationsEdit - Netscape
File Edit View Go Bookmarks Tools Window Help
Search
Home
Netscape
Bookmarks
What's inside
Ask Us!
Selected library resources & services
Mobile Librarian home
August 2004
S M T W T F S
switch month
back
current
next
Aug
2004
view schedule
appointments
University > College > Mobile librarian
UNIVERSITYLIBRARIES
Schedule an appointment
August 17, 2004
Internet ID:
Date:
8/17/2004
Start time:
End time:
Description:
Sample Appointment
Appointment location: (select one)
My office
Room 309
Magrath Library (St. Paul)
Wilson Library (West Bank)
Computer available:
yes
no
Librarian:
Status:
active
Question and response
Your question:
Document: Done

http://www.lib.ncsu.edu/risd/staff/sneilson/

NCSU LIBRARIES

Home | Services | Research Resources | About the Libraries | NC State

Search Catalog | Database Finder | E-journal Finder |

Research and Information Services

Faculty Services for classes

Subject Guides

Class Pages

Susan Wolf Neilson home

Contact Reference Librarian

Susan Wolf Neilson

Reference Librarian for Management and Economics
Email: susan_neilson@ncsu.edu
Box 7111 NCSU
Raleigh, NC 27695-7111
Phone: 919.515.5699
Fax: 919.515.8264

Office Hours:
College of Management Nelson Hall Room 2152
Tuesdays 9-12 Wednesdays 3-5
OR--e-mail or call for appointment at DH Hill Library

Search Catalog | Database Finder | E-journal Finder | Ask a Librarian | Search Website | Site Index
Home | Services | Research Resources | About the Libraries | NC State

Copyright | Disclaimer

Last Modified: 2003-10-06
Questions/Comments to LibWebTeam
URL: http://www.lib.ncsu.edu/risd/staff/sneilson/

http://www.lib.utexas.edu/services/reference/libonloc.html

THE UNIVERSITY OF TEXAS AT AUSTIN

Home | Sitemap | Help

University of Texas Libraries

SEARCH: Library Web Site GO

About the Libraries | Research Tools | Library Services | Resources for You

Ask a Librarian

Home : Services - Reference - Librarian on Location

For Students:
Finding Articles
Finding Books
Finding Reserves
Citing Resources
Evaluating Information

For Faculty/ UT Staff:
Alert Services
Citation Indexes
Dissertations
E-Journals
Purchase Request
Reserves Request

For a complete listing, see Faculty/UT Staff Services

Librarian on Location

Librarians with laptops will be available to help you with your research and information needs.

Location: Jester 1st floor, Concourse West
Time: Wednesdays, noon - 4:00 pm
Dates: August 25 through December 1, 2004

If you have any questions or comments, please **email us** or call or visit any **library reference desk**.

UTDIRECT
requires EID

UTOPIA

Accessibility | Privacy/Confidentiality | Material Usage Statement
Comments | About This Site
Page viewed: November 9, 2004 | Page last modified: August 27, 2004

http://www.lib.utexas.edu/services/reference/refdir.html

THE UNIVERSITY OF TEXAS AT AUSTIN

Home | Sitemap | Help

University of Texas Libraries

SEARCH: Library Web Site GO

About the Libraries | Research Tools | Library Services | Resources for You

Ask a Librarian

Home : Services : Ask a Librarian : Call or Visit Us

Ask a Librarian: Call or Visit

Library name	Phone number	Location	Reference Hours
Architecture and Planning Library	512-495-4620	BTL 200	M-Th 9am-10pm; F 9am-5pm; Sat noon-6pm; Sun 3pm-10pm
Benson Latin American Collection	512-495-4522	SRH 1.108	M-Wed 9am-10pm; Th 9am-6pm; Fri 9am-6pm; Sat 1pm-5pm; Sun 2pm-10pm
Center for American History	512-495-4532	SRH 2.101	M-Sat 9am-5pm; closed Sun
Chemistry (Mallet) Library	512-495-4600	WEL 2.132	M-Th 8am-10pm; Fri 8am-7pm; Sat 1pm-5pm; Sun 1pm-10pm
Classics Library	512-495-4690	WAG 1	M-F 8am-5pm; closed Sat and Sun
Engineering (McKinney) Library	512-495-4511	ECJ 1.300	M-Th 8am-10pm; Fri 8am-5pm; Sat 1pm-5pm; Sun 2pm-10pm
Fine Arts Library	512-495-4480	DFA 3.200	M-Th 8am-10pm; Fri 8am-5pm; Sat 1pm-5pm; Sun 1pm-10pm
Geology (Walter) Library	512-495-4680	GEO 302	M-Th 8am-10pm; Fri 8am-6pm; Sat 1pm-5pm; Sun 2pm-10pm
The Harry Ransom Center	512-471-9119	HRC 5.202	M-Fri 9am-5pm; Sat 9am-noon
Jester Center		1st floor, Concourse West	Wednesdays noon-4pm
Life Science Library	512-495-4630	MAI 220	M-Th 8am-10pm; Fri 8am-

http://www.lib.utexas.edu/services/reference/refdir.html

			6pm; Sat 1pm-5pm; Sun noon-10pm
Marine Science Library	361-749-6778	Marine Science Institute, Port Aransas, Texas	M-Fri 8am-5pm; closed Sat and Sun
Perry-Castañeda Library (Main)	512-495-4250 1-866-895-4250 (toll free)	PCL 2.200	M-Th 8am-9:45pm; Fri 8am-8am-4:45pm; Sat noon-4:45; Sun noon-9:45
Physics-Mathematics-Astronomy (Kuehne) Library	512-495-4610	RLM 4.200	M-Th 8am-10pm; Fri 8am-5pm; Sat 1pm-5pm; Sun 2pm-10pm
Public Affairs (Wasserman) Library	512-495-4400	SRH 3.240	M-Th 8am-10pm; Fri 8am-5pm; Sat 1pm-5pm; Sun 1pm-10pm
Tarlton Law Library	512-471-6220	JON 2.200	M-Th 7:30am-midnight; Fri 7:30am-6pm; Sat 9am-6pm; Sun 10am-midnight
Undergraduate Library	512-495-4437	FAC 101	M-Th 9am-9:45pm; Fri 9am-4:45pm; Sat noon-4:45pm; Sun noon-9:45pm

UTDIRECT
requires EID

UTOPIA

Accessibility | Privacy/Confidentiality | Material Usage Statement
Comments | About This Site
Page viewed: November 4, 2004 | Page last modified: August 27, 2004

http://www.lib.washington.edu/subject/librarians/mudrock.html

UNIVERSITY OF WASHINGTON Libraries

Libraries Home | Resources | Services | About

Theresa Mudrock

Theresa Mudrock

History Librarian

Email: mudrock@u.washington.edu
Office: 136 Suzzallo Library
Phone: (206) 685-1623
Website: http://faculty.washington.edu/mudrock

As History Librarian, Theresa is responsible for supporting the research and teaching needs of the Department of History. This includes purchasing materials, providing research assistance and instructing students in research methods.

Other job responsibilities include selecting materials for Suzzallo Reference, creating and maintaining a variety of webpages, and serving as fund manager for the humanities. In addition, Theresa teaches HIST221/INFO221, Research Strategies in History.

Fall Quarter Office Hours:
Smith 210, Wednesdays 11:00 - 1:00 (begins October 6)

Selected Webpages:

- History Subject Page
- Suzzallo Reference
- Reference Tools
- History Day Research at the UW Libraries
- Interrupted Lives
- Camp Harmony

Archive of Class Help Pages

Librarians with related expertise:
Sarah Leadley, UW Bothell
Suzanne Klinger, UW Tacoma

Questions? Comments? Email: mudrock@u.washington.edu
Last modified: Wednesday October 13, 2004

http://www.lib.uwo.ca/taylor/engineering.shtml

Text Only Version

Western Libraries

Allyn & Betty Taylor Library

The Engineering, Health Sciences, Medicine & Dentistry and Science Library

ENGINEERING

Resources by Subject

- Chemical & Biochemical
- Civil & Environmental
- Electrical and Computer
- Mechanical and Materials
- Software Engineering

Commonly Used Databases

- ASTA
- CISTI Source
- Ei Compendex
- ENGnetBASE
- IEEE Xplore
- Inspec
- National Technical Information Service
- SciFinder Scholar
- Web of Science

Standards

- ANSI
- ASTM
- Canadian Standards Association
- IEEE Xplore
- ISO

Patents

- Canadian Intellectual Property Office
- European Patent Office
- Japan Patent Office
- LexisNexis Patents
- US Patent and Trademark Office

New Books

- New Engineering Books

Writing and Style Guides

- ACS Style Guides
- Style Guides
- Report Writing
- Technical Writing

UWO Departments

- Chemical & Biochemical
- Civil & Environmental
- Electrical and Computer
- Mechanical and Materials

Research and Instructional Support

Pilot Project Reports

Proposal for Liaison Outreach Services to the Faculty of Engineering

Background

The field of engineering has been a leader in the conversion of research materials to electronic format. The Science and Technology Library has more electronic reference materials in this field than in any other.

With the completion of the three new engineering buildings, the Faculty of Engineering will have completely moved to the western edge of the campus by late 2004, and away from the Science and Technology Library.

As a result of these two developments, we anticipate seeing a reduced number of our Engineering clientele in the Science and Technology Library.

Proposal

In order to maintain and build upon the good working relationship between the Science and Technology Library and the Faculty of Engineering, and to keep the profile of the University of Alberta Libraries strong in this Faculty, we propose to place engineering liaison librarians in the Faculty, on a pilot basis, to provide library services four hours per week, in two two-hour time periods. Service to begin in September 2003, and run until at least December 2003, perhaps until April 2004, at which time the service will be reviewed.

Target User Groups

Primary: Undergraduate students; Secondary: graduate students, faculty and staff

Staff

Two engineering librarians (each working two of the proposed four hours)

Space and Equipment

Computer lab or office space with an Internet connection to be provided by the Faculty of Engineering. Computer station or laptop; telephone if available.

Duties

Library and research skills instruction, consultation, information services as requested

Assessment

Statistics will be kept, identifying numbers of questions and users (are they undergrads, grads, etc.) Users will be asked to complete a follow-up questionnaire (optional). The service will be evaluated when the pilot is completed.

Preferred Outcomes for Engineering

- Increased awareness and use of online engineering resources.
- Increased awareness of services provided by the Science & Technology Library, and the engineering librarians
- Savings of time and money through increased access to and use of information resources purchased by the library
- Strengthening awareness among library staff of the changing needs of the Faculty of Engineering

Preferred Outcomes for the Science & Technology Library

- Increased visibility of the UA Libraries for all users in the Faculty of Engineering
- Increased use of the Libraries' services, particularly the online resources
- More one-on-one and small group library and research skills instruction
- Ongoing development and strengthening of liaison relationship with the Faculty

Randy Reichardt, Jerry Kowalyk and Sandy Campbell,
Science and Technology Library, University of Alberta
4 September 2003

UNIVERSITY OF ALBERTA
LIBRARIES

Thank you for using **Librarian On-Site!** today. We would appreciate your feedback on this trial service, which will run until April 2004. Your feedback and comments will help us determine whether to continue providing the service in the fall of 2004.

1. What kind of information were you looking for today?

2. Why did you choose to use the **Librarian On-Site!** service, instead of another information source?

3. The librarian on duty helped me

	strongly disagree				*strongly agree*	
a) identify resources I needed	1	2	3	4	5	N/A
b) retrieve resources I needed	1	2	3	4	5	N/A
c) evaluate the information I needed	1	2	3	4	5	N/A
d) learn how to find and retrieve information	1	2	3	4	5	N/A
e) navigate the UA Libraries' web site	1	2	3	4	5	N/A

4. The librarian who helped me was

	strongly disagree				*strongly agree*	
a) approachable and welcoming	1	2	3	4	5	N/A
b) available when I needed help	1	2	3	4	5	N/A
c) courteous and polite	1	2	3	4	5	N/A
d) knowledgeable in						
- finding general information	1	2	3	4	5	N/A
- resources in my discipline	1	2	3	4	5	N/A

5. Were you satisfied with the help you received with your question(s) today?
   a) ___ extremely
   b) ___ very
   c) ___ somewhat
   d) ___ not very
   e) ___ not at all

6. During the trial period, librarians will be available for consulting on Wednesdays and Thursdays from 1300-1500 hrs each week, in E2-006. Are you satisfied with these hours of availability?
   Yes___ No ___

Please indicate any other hours of availability you would prefer for this service:

Science and Technology Library

Cameron Library _ University of Alberta _ Edmonton _ Canada _ T6G 2J8
Telephone: (403) 492-_ Fax: (403) 492-

7. Are you satisfied with the location (Computer Lab ETLC E2-006) for this service?
Yes ___ No ___

If no, please indicate where you would like to see the service offered within the Faculty of Engineering.

8. In what ways do you think the **Librarian On-Site!** service could be improved?

9. If you have used other types of information and reference services on campus, please indicate which services they are:

a) __ in-person reference services at a subject library
b) __ online chat reference
c) __ e-mail reference
d) __ telephone reference
e) __ private appointment with a subject librarian
f) ___ other, please specify: ____________________

10. Please estimate how many times you have used the library during this current school term
a) ___ daily
b) ___ several times a week
c) ___ once a week
d) ___ 2-4x a month
e) ___ other, please specify: ____

11. What best describes you?
a) ___ undergrad student; Year of study ___
b) ___ masters student
c) ___ PhD student
d) ___ researcher (PDF, etc.)
e) ___ faculty member
f) ___ staff member

12. What is your current subject discipline?
a) ___ Chemical and Materials
b) ___ Civil and Environmental
c) ___ Electrical and Computer
d) ___ Engineering Management
e) ___ Mechanical
f) ___ Mining and Petroleum
g) ___ NINT
h) ___ Other, please specify ______________________________

13. Would you like to see the **Librarian On-Site!** service continued in 2004? Why or why not?

Thank you for taking the time to complete our questionnaire. Please return to:

Randy Reichardt, Science and Technology Library, 1-26 Cameron Library

UNIVERSITY OF ALBERTA
LIBRARIES

Thank you for using **Librarian On-Site!** today. We would appreciate your feedback on this trial service, which will run until April 2004. Your feedback and comments will help us determine whether to continue providing the service in the fall of 2004.

1. What kind of information were you looking for today?

Reference data (i.e. density, ash content of bone)
Airflow/fluid mechanics
Nanotube info
Supplementary references to an article
How to locate sources found on the databases
Turboexpanders
Historical research for a HIST 391 essay (ITS elective)
Heat fin transfer theory
Nanotechnology applications & carbon nanotubes
How to find review papers
Journal article location + author location
Info for a technical report about LEED, innovative design and any other interesting topic
Accessing a hard copy of an online gov't report
Finding some journals about my area of project
Coke gasification; gasifier & sulfur removal
Biotechnology
Stuff on turbochargers and hydraulics

2. Why did you choose to use the Librarian On-Site! service, instead of another information source?

I had already tried searching to the best of my ability
Did not know where to look
Convenience
Could not find the info I needed through the web-site
Close to where I was
More helpful
Easy access; good environment
Right there and needed to do it!
Because it is more reachable + a lecture by a librarian in my course lectures
It is easy access, good information specific answer to my question as I need. Person interaction is good.
Because I learn best with one-on-one, and then I also remember it better for next time
Convenience, I needed to ask a librarian and the librarian-on-site was the closest one
I can be helped more and I can ask my questions
Randy is knowedgable & professors recommended him
It was very handy to have the help in the computer lab as I was looking up info
He was there

Science and Technology Library

Cameron Library _ University of Alberta _ Edmonton _ Canada _ T6G 2J8
Telephone: (403) 492-_ Fax: (403) 492-

3. The librarian on duty helped me

	strongly disagree *1*	*2*	*3*	*4*	*5* *strongly agree*	*N/A*
a) identify resources I needed	0	0	0	3	17	2
b) retrieve resources I needed	0	2	0	4	16	1
c) evaluate the information I needed	0	0	3	2	9	7
d) learn how to find and retrieve information	0	0	0	4	17	1
e) navigate the UA Libraries' web site	0	0	1	5	15	1

4. The librarian who helped me was

	strongly disagree *1*	*2*	*3*	*4*	*5* *strongly agree*	*N/A*
a) approachable and welcoming	0	0	0	0	22	0
b) available when I needed help	0	1	0	1	20	0
c) courteous and polite	0	0	0	0	22	0
d) knowledgeable in						
- finding general information	0	0	0	1	20	1
- resources in my discipline	0	0	0	1	20	1

Definitely! [from 4c]

5. Were you satisfied with the help you received with your question(s) today?

a) _14_ extremely
b) _7_ very
c) _1_ somewhat
d) _0_ not very
e) _0_ not at all

6. During the trial period, librarians will be available for consulting on Wednesdays and Thursdays from 1300-1500 hrs each week, in E2-006. Are you satisfied with these hours of availability?

Yes _17_ No _5_

Please indicate any other hours of availability you would prefer for this service:

5 days/week
Every day for 1 hr morning 1 hr afternoon?
I would like even more availibility
Longer availability (1-5 pm during busy times)
M, T, F → it would be good to have constant access
Mondays @ same time
More please! Some in the morning would be cool

7. Are you satisfied with the location (Computer Lab ETLC E2-006) for this service?

Yes _21_ No _0_

If no, please indicate where you would like to see the service offered within the Faculty of Engineering.

8. In what ways do you think the Librarian On-Site! service could be improved?

I was not able to get the article I requested, only because the library did not have the journal I needed and due to time constraints it could not be brought in
Librarian on-site service is excellent
No lines on computer; new mouse!
I think this way is good because we can ask our questions and get help
I am aware of many students who do not know how to use the databases. I know word of mouth spreads this info about this service but a little advertisement of this service would be nice for students who do not know of this service.
It is very good right now
By having lectures in other courses like MATE 353 and by giving student assignments…
More hours
More hours
More days

9. If you have used other types of information and reference services on campus, please indicate which services they are:

a) _13_ in-person reference services at a subject library
b) _4_ online chat reference
c) _3_ e-mail reference
d) _2_ telephone reference
e) _3_ private appointment with a subject librarian
f) _0_ other, please specify: ____________________________

10. Please estimate how many times you have used the library during this current school term

a) _3_ daily
b) _11_ several times a week
c) _2_ once a week
d) _4_ 2-4x a month
e) _1_ other, please specify: _2-4x a term, but expect to increase further into Eng. degree_

11. What best describes you?

a) _19_ undergrad student; Year of study _2nd – 1; 3rd – 3; 4th – 12; 5th - 3_
b) _3_ masters student
c) _0_ PhD student
d) _0_ researcher (PDF, etc.)
e) _0_ faculty member
f) _0_ staff member

12. What is your current subject discipline?

a) _8_ Chemical and Materials
b) _3_ Civil and Environmental
c) _5_ Electrical and Computer
d) _1_ Engineering Management
e) _6_ Mechanical
f) _0_ Mining and Petroleum
g) _0_ NINT
h) _0_ Other, please specify ____

13. Would you like to see the Librarian On-Site! service continued in 2004? Why or why not?

Yes, my design project is next semester and I will need even more help!
Yes – it makes accessing information much easier! I really appreciated the help I received this year (on 4-5 different occasions).
Yes. This service is excellent. It makes finding information easier than going to Cameron Library. I wish I would have had access to this service in my earlier years as an undergrad.
Yes! It is close by.
Yes. It's helpful and we don't have to go to Cameron.
Yes! Very helpful!
Yes, it helps students find what they are looking for
Yes I think it is a good way to help students
Yes, the service is very convenient and helps students make use of library services and resources that are not familiar to them
Yes, because there is always something new to learn and there is always somewhere else to find more information.
Yes.
Yes because it is the help students need to find information for school projects
Of course, because it helped me in looking for articles & database about my project
Yes, it is extremely convenient and good for students who are unfamiliar with the library
Yes! It makes it accessable to us and we are more likely to learn about resources the library has to offer.
Yes.
Yes. It is a very valuable resource that remains untapped by many students. The more who know about it, the more need there will be for it.
Absolutely, it's nice not to have to go to Cameron since we are almost all here now.
Yes – it is a great resource.
Yes! It is very handy and helpful. Most engineering work is done in ETLC at the computers and it is great to have such accesible references.

Thank you for taking the time to complete our questionnaire. Please return to:

Randy Reichardt, Science and Technology Library, 1-26 Cameron Library

http://www.rci.rutgers.edu/~kuchi/Web_version/L_W_B_Report.html

Librarians without Borders

Report of the Pilot project
of the Outpost Services Team

June 2002

***Pilot Location*:**	**Busch Campus Center Information Desk**
***Pilot Period*:**	**April 8 – May 8, 2002 Monday - Thursday**
***Pilot Hours*:**	**11:00 a.m. – 2:00 p.m.** **7:00 p.m. – 9:00 p.m.**

Outpost Services Team: A task force appointed by the New Brunswick Libraries' Information Services Steering Group

Members:

Triveni Kuchi, Social Sciences/Instructional Services Librarian.
Laura Bowering Mullen, Librarian.
Stephanie Tama-Bartels, Instructional Services Librarian.

http://www.rci.rutgers.edu/~kuchi/Web_version/L_W_B_Report.html

Librarians without Borders

Report of the Pilot project
of the Outpost Services Team

Contents

http://www.rci.rutgers.edu/~kuchi/Web_version/L_W_B_Report.html

Introduction

As part of its strategic planning process, the New Brunswick Libraries Information Services Steering Group's (NBL ISSG) annual report 2000-2001 put forth the following objective: "continue to explore innovative ways of offering information/reference services."

In spring semester, 2002, NBL ISSG created an Outpost Services Team[1] to develop a framework to study the demand for library services at locations other than library buildings. This framework for outpost services would:

- Explore general issues involved in such library outreach programs
- Investigate the nature of the demand for library services beyond the physical space of the libraries
- Survey the policies and procedures required to implement such programs
- Investigate the best methods for providing such programs
- Create an implementation plan for the programs
- Evaluate the type and nature of services requested by the user
- Conduct and analyze user-response surveys
- Explore the long term feasibility of such programs

Since this was the first time the Libraries were looking at setting up an outpost program of this nature, not much background information was available. A literature review was not of much help either. NBL ISSG identified in its annual report ways to experiment with outreach services, "such as establishing library outposts in dorms or student centers, and establishing library cyber-posts". It suggested that the Outpost Services Team initially set up a pilot program at Campus/Student centers during 2002-spring semester.

With further discussions, the team explored types of library services and locations that would be suitable for an outpost. Experimenting with a pilot program in campus centers seemed to be good ways to gather more information and to further understand the needs of an outpost.

Goals for the Pilot Program

- Create a satellite outpost as an additional point of entry for library services
- Provide library services where students congregate on campus
- Offer this service during "research/term paper" time in the semester
- Test and evaluate the need for and interest in such services

[1] Members of the Team are Triveni Kuchi, Laura Bowering Mullen, and Stephanie Tama-Bartels.

http://www.rci.rutgers.edu/~kuchi/Web_version/L_W_B_Report.html

Planning the Pilot Outpost Service

Early in the planning process, a meeting was arranged with the associate dean of Rutgers College and the director of the Rutgers College Student Centers.[2] They enthusiastically responded to the idea of an outpost, which, from their perspective, added to the breadth of services offered at the campus centers. Working closely with the university administration was imperative to the implementation and success of the pilot program.

The team met several times and discussed various issues of location, cost, time for set up, hours of service, materials and feasibility. In its initial discussions, the team determined that, in general, reference and access services, handouts and library promotional materials could be provided at a library outpost. The type of services to be offered depended largely on the physical setup available at a particular location. If the physical set up included a network connection, the outpost could provide reference and access services, along with informational and promotional materials such as handouts, brochures and other advertising. If no network connection were available, only informational and promotional materials would be possible.

The locations under consideration were the Rutgers College Student Centers, which include the Rutgers Student Center on College Avenue in New Brunswick and the Busch Campus Center on the Busch Campus in Piscataway. Consultations with the Director of the Rutgers College Student Centers and Associate Director of the Busch Campus Center (BCC) helped determine which campus center would be suitable for these different types of library services. The determinative factors for selection of a location were feasibility, cost, and time for setting up an outpost. Both campus centers had facilities available for an outpost, though BCC was more readily available and could be adapted very quickly with minimal cost as compared to College Avenue Student Center. Therefore, the team decided that the BCC would be an ideal location for the pilot program.

Selection of the Location

The information desk at BCC is located centrally to the three entrances into the building. The Campus Center Administration generously offered one half of this circular information desk for the library outpost. Centrality of the desk and over 15000 visitors per day gave the outpost high visibility and provided a convenient place for students, faculty and public to stop by and ask questions. In addition, computers with email capabilities, immediately adjacent to the information desk, had the potential to draw indirect traffic to the library outpost.

The information desk also could be conveniently used as a library reference desk without major changes or construction. There was enough space to place two computers, connections to the network, and a printer. In addition, it came with some security that permitted us to leave the computer equipment and library materials in place without having to dismantle and set up.

[2] Rutgers College is selected for study since it is the largest of the New Brunswick/Piscataway colleges.

The Team had only two weeks to plan and prepare the outpost to meet the objective of providing the service during the "research/term paper" time in the semester. The information desk was ideal since it was "ready-to-go" and did not require extensive set up.

There was almost no budget to cover costs for the project and, while there were rooms within the campus center that had fees associated with them, the space at the information desk was offered free of charge for the library outpost. So, the decision about location was based on high visibility, convenience, time for preparation, and cost.

Selection of the Pilot Period

Once the location was selected, the first step was to decide the time frame for the pilot. April 8 through May 8, 2002 (between Spring break and exams) seemed most suitable, as this was the "research/term paper" time in the semester. It is expected that students need more help with library research when they are writing their term papers. This also provided sufficient time to study and gather data.

Selection of Hours/Days for the Pilot

According to statistics provided by BCC, the "peak" hours for foot traffic in the building were 11:00 a.m. to 2:00 p.m. and 8:00 p.m. to 11:00 p.m., Monday through Thursday. Based on this information, the Team decided to keep the outpost open these four days for all of the peak morning hours but, due to staffing considerations, evening hours were more realistically set from 7:00 p.m. to 9:00 p.m.

Scheduling staff at the Outpost for the Pilot

Initially we decided to staff the outpost with volunteer librarians. However, given the time required to solicit and schedule volunteers and since the pilot would run during the busiest time of the semester for in-library reference, the team decided to limit staffing to its three members. This eliminated the need for training the volunteer librarians and enabled the team to collect consistent, accurate data and make notes based on direct observations.

For the first week, two team members were scheduled to work each shift. After initial observations of the number of transactions, it was decided that, for the rest of the pilot period, the outpost could be covered by only one team member unless there was an increase in activity. Scheduling was done one week in advance, taking into consideration each member's calendar and observations from the previous week.

http://www.rci.rutgers.edu/~kuchi/Web_version/L_W_B_Report.html

Materials for the Pilot

A networked computer with a printer was set up at the outpost to handle reference questions that could be answered with digital resources accessible from the RUL Website and the World Wide Web. The team members brainstormed about other materials and determined that the following were needed:

- Handouts on library services, remote access, borrowing privileges, hours, New Brunswick library locations and map, Web evaluation tips, etc.
- Display racks for handouts and signs
- Office supplies
- Box to collect surveys
- Name badges

Advertising for the Pilot

The target audiences for advertising this service were Rutgers students and faculty members. In order to reach out to them, a memo was sent to the Associate Dean of Rutgers College who in turn informed undergraduate program directors and faculty advisors to encourage students to use the service. Several venues were chosen to advertise services directly to the students. Flyers along with the dates and time were posted at BCC as well as dormitories. (Exhibit "A") A banner with the Rutgers Libraries' logo and URL was displayed across the front of the outpost desk. A poster-sized version of the "Librarians without Borders" flyer was placed at the main entrance to the BCC. Another poster entitled "Librarians can help you..." was set up alongside the outpost listing a few examples of how librarians could provide assistance. (Exhibit "B") Finally, a press release was drafted for the Targum and for the "News & Events" feature on the Libraries' Website. (Exhibit "C")

User Surveys and Statistics for the Pilot

To test and evaluate the need for and interest in outpost services, a brief survey and statistics were collected. The survey posed three questions that would provide data for evaluating whether the service was helpful, if the patron could use the service in the future and possible additional locations at which it could be provided. The surveys were left visibly on the desk for anyone to complete, but one was specifically handed to the patron after a reference transaction. Collection sheets for gathering statistics recorded the number of transactions at the desk as well as the type (reference or directional), duration, day or evening shift and subject area of each transaction.

http://www.rci.rutgers.edu/~kuchi/Web_version/L_W_B_Report.html

Report on Statistics

The pilot outpost service was provided for 19 days Monday through Thursday for a total of 95 hours (5 hours per day - 3 hours in the day and 2 hours in the evening).

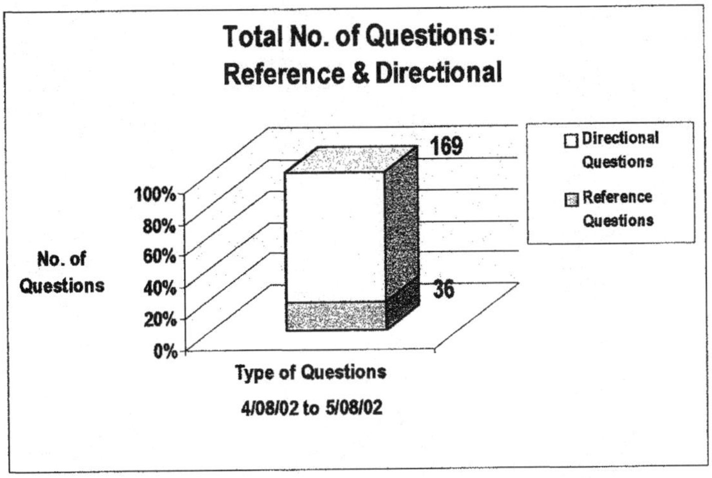

It is interesting to note that, of a total of 205 questions answered during this period, 169 (82%) constituted directional questions and 36 (18%) were reference.

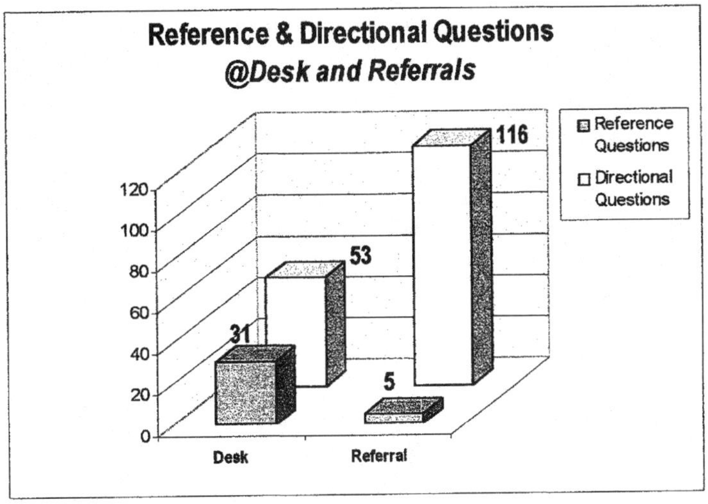

The high number of directional questions reflects the proximity of the outpost to the BCC information desk. Of the directional questions, 53 (31%) were answered at the desk and 116 (69%) were referred to the BCC information desk assistants. Examples of directional

http://www.rci.rutgers.edu/~kuchi/Web_version/L_W_B_Report.html

questions handled by librarians at the outpost included bookstore hours, directions to Administrative Services Building, requests for supplies and college newspapers. The directional questions referred to the BCC information desk assistant normally pertained to BCC activities, such as the location of an event, and assistance with equipment (copy machine, television).

Of all the reference questions, 31 (86%) were answered at the desk and 5 (14%) were referred to other Rutgers Libraries. The reference questions could be answered at the desk using the digital resources available. Team members assisted patrons with library research on various topics such as Niebuhr theories of democracy, ethnology, folk law and anthropology, x-ray diffraction, literature reviews in art history, and construction management. While the number of reference questions asked was few, almost all entailed in-depth research assistance. There were very few quick reference or access services related questions. Questions that were referred to libraries included directing the patron to the Library of Science and Medicine in order to obtain a 1991 citation, sending the patron to Libraries administration for assistance, etc.

Comparing Questions asked during day and evening shift

	Totals	A.M.11:00-2:00	P.M.7:00-9:00
Reference Questions	36	25	11
Directional Questions	169	114	55
All Questions	205	139	66

68% of the 205 questions were asked in the day. Of these questions, 12% were reference and 56% were directional. Of the 32% asked in the evening, 5% were reference and 27% were directional.

It can be seen from above that the service was used more often during the day. This information would be useful for scheduling staff at an outpost in the future. However, it is important to note that the outpost was open for 3 hours in the day as compared to 2 hours in the evening, which might be the cause for more questions being asked during the day shift. Collecting "hourly" statistics for future projects would enable additional manipulation of data and help clarify staffing needs more precisely.

http://www.rci.rutgers.edu/~kuchi/Web_version/L_W_B_Report.html

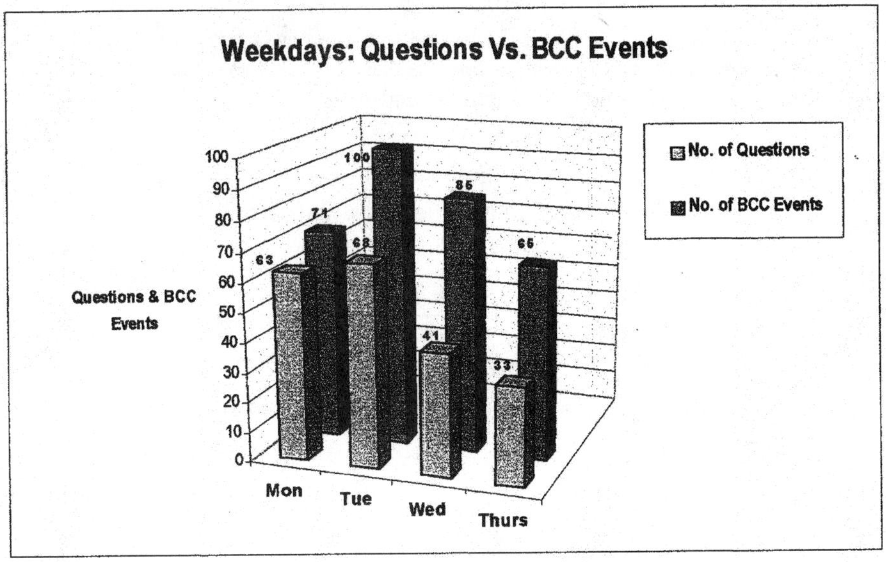

Of the four weekdays in which the outpost operated, the most questions were asked on Tuesdays, closely followed by Mondays. In response to a request for BCC traffic statistics, the Associate Director of the center provided a list of events held at the center for each day of the pilot period. This data solely reflected the registered events and not attendance. Yet, it gave preliminary information to compare number of events to the number of questions asked at the outpost. Interestingly, Tuesdays had the highest number of registered events as well as the highest number of questions. However, the rest of the data was too general to support a trend between BCC events and questions.

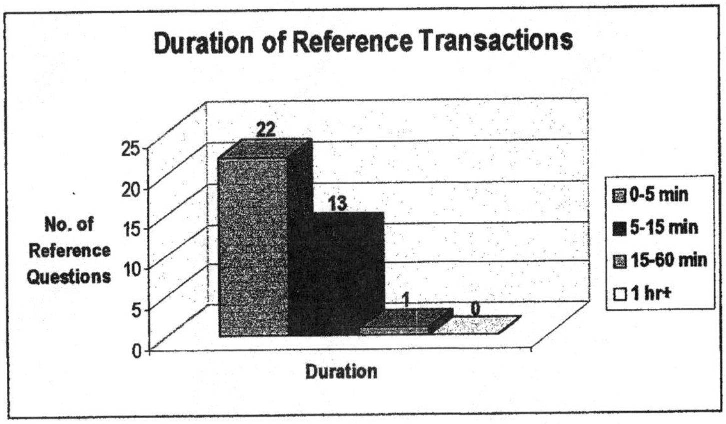

http://www.rci.rutgers.edu/~kuchi/Web_version/L_W_B_Report.html

61% of the reference questions were answered in 5 minutes or less; 36% lasted between 5 and 15 minutes; more than 15 minutes but less than an hour was spent on 3% of the questions; and there were no transactions beyond an hour.

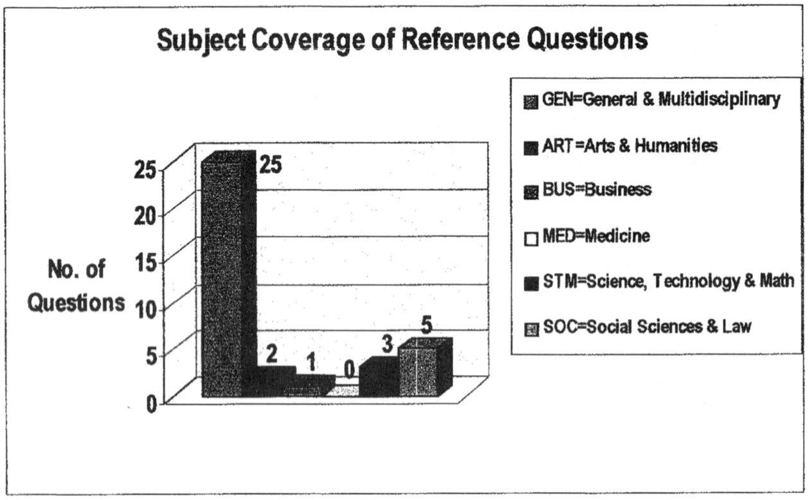

69% (25) of reference questions asked were general or multidisciplinary in nature, such as borrowing privileges for graduate students, finding a chemistry class syllabus online, and locating and printing the Rutgers University historical timeline. 14% (5) of the questions were related to social sciences and law. Only 8% of reference questions related to science, technology and math. This was surprising since BCC is on a science campus and more of these types of questions were expected.

http://www.rci.rutgers.edu/~kuchi/Web_version/L_W_B_Report.html

Report on User Surveys

There were 10 respondents to the user surveys handed out by the team members after the reference transaction. Of the 31 reference questions asked at the desk, this was a response of only 32%. 100% responded positively to the question "Did you find the service useful?" Some remarked that it is "excellent," "convenient", "extremely [useful]", "[an] easier, more convenient location", and "[the librarians] went out of their way with directions and assistance in a friendly and open manner". When asked, "Would you like to use the service in the future?" 90% said yes, and included comments such as "for sure" and "glad it was centrally placed". "Where else do you think this service can be offered?" was answered by 60% of the survey respondents. Suggestions for potential locations included "all student centers" and "computer labs".

It is important to note that the 10 responses received were all the result of a team member asking the patron to fill out a survey. There was no instance of a user filling out a survey without prompting. The low turnout of the survey responses could be attributed to the patron choosing not to respond or the survey not being handed out in time for the patron to respond. The team needs to further evaluate how to administer a more efficient and effective survey at an outpost.

Conclusion and Recommendations

Partnership with university administration was crucial to the implementation and success of the program. It not only provided the facilities for the outpost, but also helped promote it. Given such a set up, including a network connection, the outpost was able to function as a reference desk as well as for library promotion. Two hundred five questions were asked over the course of the 19 days of the pilot. Of these, the majority (82%) were directional questions, many of which were referred to the BCC information desk assistant. The reference questions were only 18% of the total.

In light of the low turnout of reference questions at the BCC, the decision to make this outpost permanent first warrants a user study that would look at the needs and expectations of the user for library services at various locations including campus centers and computer labs. The results of the study may suggest services that we have not yet conceptualized or offered.

The team promoted the outpost to the targeted audience in the limited time available (two weeks), which was not sufficient time for the Rutgers community to become aware and use the service. Advertising further in advance of setting up the outpost and using more creative and aggressive marketing techniques might have had more impact on the use of the service. One means of publicity would be opening the BCC outpost during Rutgers College orientation and for the first week of the Fall, 2002, semester. Beyond providing publicity for the outpost, this would help test it as an informational and promotional type of service and send the message to new students to expect library services outside of the libraries.

http://www.rci.rutgers.edu/~kuchi/Web_version/L_W_B_Report.html

LIBRARIANS WITHOUT BORDERS

Need help?

Ask a Librarian at the

Busch Campus Center

April 8 - May 8
Monday - Thursday
11 am–2 pm & 7 pm–9 pm

Rutgers University Libraries

http://www.rci.rutgers.edu/~kuchi/Web_version/L_W_B_Report.html

LIBRARIANS CAN HELP YOU...*

- Find information on just about anything
- Locate books and journal articles on a topic
- Find and evaluate online information

* These are just a few examples

Rutgers University Libraries

http://www.rci.rutgers.edu/~kuchi/Web_version/L_W_B_Report.html

Rutgers University Libraries

News and Events:
Archive:

Pilot Project: Librarians Without Borders, at Busch Campus Center

Libraries Home
Ask a Librarian
Hours and Directions
Search Website
Site Index
Advanced Users
IRIS
Indexes and Databases
Electronic Journals
Electronic Books
Subject Research Guides
Electronic Reference Sources
Other Library Catalogs
Digital Library Projects
How Do I...?
User Services
Library Instruction and Tutorials
Off-Campus Support
News and Events
About the Libraries
Friends and Supporters
Staff Resources

In an effort to provide research services and support to students beyond library settings, the R University Libraries in New Brunswick have set up a virtual library outpost at the Busch Campu (BCC) as a pilot project.

This new service builds upon the notion of a library 'without walls', a phenomena of increasin our Internet-enabled world. Librarians participating in this pilot project will provide students wit how to navigate the complex world of the digital library using electronic journals, databases, V and electronic books. When appropriate, students will be referred to print and other resource libraries.

This pilot program, Librarians Without Borders, has been developed and implemented by the Services Team appointed by the Rutgers University New Brunswick Libraries Information Serv Group. This service is available at the Information Desk in the BCC, from April 8th through Ma through Thursday from 11am-2pm and 7pm-9pm.

For your comments and input on this new service, please contact the members of the Outpos Team:

Stephanie Tama-Bartels
stama@rci.rutgers.edu
732/932-7129 ext.131

Triveni Kuchi
kuchi@rci.rutgers.edu
732/445-5733 #2

Laura Bowering Mullen
lbmullen@rci.rutgers.edu
732/445-4806

RUTGERS
Return to RU Main Menu

Posted April 19, 2002

URL:
http://www.libraries.rutgers.edu/rul/news/archive_02/2002_04_libs_w-out_bc
Libraries website maintained by the Libraries Webmaster

http://www.rci.rutgers.edu/~kuchi/Web_version/L_W_B_Report.html

Miscellaneous Comments by Patrons/Team Members

Two people from the board of governors (we think) meeting were very interested in the service, asked questions like "will you be there at the other student centers?" and "what services are you attempting to provide?"

"You don't look like normal librarians".

Trying to look friendly and more casual; we thought we would attract more people. Is it possible that people trust more a librarian that looks "like a librarian" - any research on this?

Two long-time science professors stopped by to ask what kind of help we expect to give students here. When I told them, they said they thought it was a wonderful idea - really liked it - many compliments.

People that stopped seemed to stop again -- more advertising might help since people may be more likely to stop if they are already familiar with the existence of the service. Also sprucing up the posters and signs would help.

Info desks need to be more aware of the library outpost service and need training to refer to outpost.

"Is this an event? Why are you here? Because I see you out here every day and I was beginning to wonder".

Some patron wanted to tell us that both she and her friend who asked us here to help with their communications papers both got A's. Very appreciative.

Person whom we helped earlier stopped by and asked how things were going.

One or two people immediately realized that we weren't the info desk, said so, and walked around to the other side of the desk to the student at the info desk.

"The library desk appeared as an oasis in a barren/ or catacombed web. Glad it was centrally placed".

"This [BCC] is perfect - very convenient when I am mulling over a problem on the way to lunch".

http://www.rci.rutgers.edu/~kuchi/Web_version/L_W_B_Report.html

Rutgers University Libraries

OUTPOSTS SERVICES TEAM

"Librarians without Borders"

Subject Codes

GEN	General & Multidisciplinary
ART	Arts & Humanities
BUS	Business
MED	Medicine
STM	Science, Technology & Math
SOC	Social Sciences & Law

11:00a.m. - 2:00 p.m. Pilot program @ Busch Campus Center

Date: **April 8th 2002 - May 8th 2002**

Transactions	0-5 min.	5-15 min.	15-60 min.	1 hour +
Reference				
Desk				
Referral				
Directional				
Desk				
Referral				

Initials & Date:

Remarks:

http://www.rci.rutgers.edu/~kuchi/Web_version/L_W_B_Report.html

Rutgers University Libraries

Subject Codes

GEN	General & Multidisciplinary
ART	Arts & Humanities
BUS	Business
MED	Medicine
STM	Science, Technology & Math
SOC	Social Sciences & Law

OUTPOSTS SERVICES TEAM

"Librarians without Borders"

7:00 - 9:00 p.m.

Date:

Pilot program @ Busch Campus Center

April 8th 2002 - May 8th 2002

Transactions	0-5 min.	5-15 min.	15-60 min.	1 hour +
Reference Desk				
Referral				
Directional Desk				
Referral				

Initials & Date:
Remarks:

http://www.rci.rutgers.edu/~kuchi/Web_version/L_W_B_Report.html

We want your Opinion Please!

Did you find this service helpful?

- ☐ Yes ☐ No

Remarks: ____________________

Would you like to use service in the future?

- ☐ Yes ☐ No

Remarks: ____________________

Where else do you think this service can be offered?

____

____

Thank You!
Librarians Without Borders

SELECTED RESOURCES

DOCUMENTS

Journal Articles

Albanese, Andrew, and Norman Oder. "Web Has Changed Campus Library." *Library Journal* 127, no. 10 (June 1, 2002): 20–22.

Blewett, Daniel K. "The Librarian Is In: Bibliographer Keeping Office Hours in an Academic Department at Loyola University." *College & Research Libraries News* no. 10 (November 1995): 701–03.

Carlson, Scott. "The Deserted Library: As Students Work Online, Reading Rooms Empty Out—Leading Some Campuses to Add Starbucks." *Chronicle of Higher Education* (November 16, 2001): A35–A38.

Davis, Stephanie R. "High Tech, High Touch: Providing Personalized Service on Users' Turf." *Behavioral & Social Sciences Librarian* 21, no. 1 (2002): 51–58.

Glynn, Tom, and Connie Wu. "New Roles and Opportunities for Academic Library Liaisons: A Survey and Recommendations." *Reference Services Review* 31, no. 2 (2003): 122–28.

Haines, Annette. "Out in Left Field: The Benefits of Field Librarianship for Studio Arts Programs." *Art Documentation* 23, no. 1 (Spring 2004): 18–20.

Hue, Terence K. "Casting a Wider Net with Roving Reference." *Computers in Libraries* 23, no. 3 (March 2004): 34–36.

Kempcke, Ken. "The Art of War for Librarians: Academic Culture, Curriculum Reform, and Wisdom from Sun Tzu." *portal: Libraries and the Academy* 2, no. 4 (2002): 529–51.

Leonard, W. Patrick. "On My Mind: Libraries without Walls—Field Service Librarianship." *Journal of Academic Librarianship* 20 (March 1994): 29–30.

Reichardt, Randy, and Jerry Kowalyk. "Librarian On-Site Service at the University of Alberta: How the Science and Technology Library Took Its Information and Reference Service from the Library to the Faculty." *SciTech News* 58, no. 3 (August 2004): 7–11.

Seamans, Nancy, and Paul Metz. "Virginia Tech's Innovative College Librarian Program." *College & Research Libraries* 63, no. 4 (July 2002): 324–32.

Schillie, Jane E., et al. "Outreach through the College Librarian Program at Virginia Tech." *Reference Librarian* no. 71 (2000): 71–78.

Tennant, Michele R. "Development and Assessment of Specialized Liaison Librarian Services: Clinical vs. Basic Science in a Veterinary Medicine Setting." *Medical Reference Services Quarterly* 21, no. 2 (2002): 21–37.

Wong, Sandra, and Anne O'Shea. "Librarians Have Left the Building: Ask Us HERE! At Simon Fraser University." *Feliciter* 50, no. 3 (2004): 90–92.

SPEC Kit Title List

SP285 Lib Svcs in Non-Library Spaces
SP284 Security in Special Collections
SP283 Grant Coordination
SP282 Managing Electronic Resources
SP281 The Information Commons
SP280 Library User Surveys
SP279 Evaluating Library Instruction
SP278 Library Patron Privacy
SP277 Lib. Pub. Access Workstation Auth
SP276 Recruitment and Retention
SP275 Laptop Computer Services
SP274 Data Mining & Data Warehousing
SP273 Chat Reference
SP272 Insuring & Valuing Res Lib Coll
SP271 Lib Systems Office Organization
SP270 Core Competencies
SP269 Integrating Preservation Activities
SP268 Reference Statistics
SP267 User Authentication
SP266 Staffing the Library Website
SP265 Instructional Support Services
SP264 Extended Library Hours
SP263 Numeric Data Services
SP262 Preservation & Digitization
SP261 Post-Tenure Review
SP260 Interview Process
SP259 Fee-based Services
SP258 Corporate Annual Reports
SP257 MLS Hiring Requirement
SP256 Changing Roles of Lib Profs
SP255 Branch Libs/Discrete Collectns
SP254 Managing Printing Services
SP253 Networked Info Services
SP252 Supprt Staff Classifictn Studies
SP251 Electronic Reference Service
SP250 TL10: Educating Faculty
SP249 Catalogng of Resrces Digitized
SP248 Licensing of Electronic Prodcts
SP247 Management of Lib Security
SP246 Web Page Devel & Managmnt
SP245 Electronic Reserves Operations
SP244 TL 9: Renovation & Reconfigur.
SP243 TL 8: Users with Disabilities
SP242 Library Storage Facilities
SP241 Gifts and Exchange Function
SP240 Marketing and PR Activities
SP239 Mentoring Programs in ARL
SP238 ARL GIS Literacy Project
SP237 Managing Food and Drink
SP236 TL 7: E-Theses/Dissertations
SP235 Collaborative Coll Management
SP234 TL 6: Distance Learning
SP233 ARL in Extension/Outreach
SP232 Use of Teams in ARL
SP231 Cust Service Programs in ARL
SP230 Affirmative Action in ARL
SP229 Evaluating Acad Libr Dirs
SP228 TL 5: Preserving Digital Info
SP227 Org of Doc Coll & Svcs
SP226 TL 4: After the User Survey
SP225 Partnerships Program
SP224 Staff Training & Development
SP223 TL 3: Electronic Scholarly Pubn
SP222 Electronic Resource Sharing
SP221 Evol & Status of Approval Plans
SP220 Internet Training
SP219 TL 2: Geographic Info Systems
SP218 Info Technology Policies
SP217 TL 1: Electronic Reserves
SP216 Role of Libs in Distance Ed
SP215 Reorg & Restructuring
SP214 Digit Tech for Preservation
SP213 Tech Svcs Workstations
SP212 Non-Librarian Professionals
SP211 Library Systems Office Org
SP210 Strategic Planning
SP209 Library Photocopy Operations
SP208 Effective Library Signage
SP207 Org of Collection Develop
SP206 Faculty Organizations
SP205 User Surveys in ARL Libs
SP204 Uses of Doc Delivery Svcs
SP203 Reference Svc Policies
SP202 E-journals/Issues & Trends
SP201 E-journals/Pol & Proced
SP200 2001: A Space Reality
SP199 Video Collect & Multimedia
SP198 Automating Preserv Mgt
SP197 Benefits/Professional Staff
SP196 Quality Improve Programs
SP195 Co-op Strategies in Foreign Acqs
SP194 Librarian Job Descriptions
SP193 Lib Develop & Fundraising
SP192 Unpub Matls/Libs, Fair Use
SP191 Prov Pub Svcs Remote User
SP190 Chang Role of Book Repair
SP189 Liaison Svcs in ARL Libs
SP188 Intern, Residency & Fellow
SP187 ILL Trends/Staff & Organ
SP186 Virtual Library
SP185 System Migration
SP184 ILL Trends/Access
SP183 Provision of Comp Print Cap
SP182 Academic Status for Libns
SP181 Perf Appr of Collect Dev Libn
SP180 Flexible Work Arrangemts
SP179 Access Services Org & Mgt
SP178 Insuring Lib Colls & Bldgs
SP177 Salary Setting Policies
SP176 Svcs for Persons w/Disabilities
SP175 Scholarly Info Centrs
SP174 Expert Systems
SP173 Staff Recognition Awards
SP172 Information Desks
SP171 Training of Tech Svc Staff
SP170 Organization Charts
SP169 Mgt of CD-ROM
SP168 Student Employment
SP167 Minority Recruitment
SP166 Materials Budgets
SP165 Cultural Diversity
SP164 Remote Storage
SP163 Affirmative Action
SP162 Audiovisual Policies
SP161 Travel Policies
SP160 Preservation Org & Staff
SP159 Admin of Lib Computer Files
SP158 Strategic Plans
SP157 Fee-based Services
SP156 Automating Authority Control
SP155 Visiting Scholars/Access
SP154 Online Biblio Search
SP153 Use of Mgt Statistics
SP152 Brittle Books Program
SP151 Qualitative Collect Analysis
SP150 Bldg Security & Personal Safety
SP149 Electronic Mail
SP148 User Surveys
SP147 Serials Control/Deselection
SP146 Lib Dev Fund Raising Capabilit
SP145 Lib Publications Programs
SP144 Building Use Policies
SP143 Search Proced Sr LibAdmin
SP142 Remote Access Online Cats
SP141 Approval Plans
SP140 Performance Appraisal
SP139 Performance Eval: Ref Svcs
SP138 University Copyright
SP137 Preservation Guidelines
SP136 Managing Copy Cataloging
SP135 Job Analysis
SP134 Planning Mgt Statistics
SP133 Opt Disks: Storage & Access
SP132 Library-Scholar Communication
SP131 Coll Dev Organization
SP130 Retrospective Conversion
SP129 Organization Charts
SP128 Systems File Organization
SP127 Interlibrary Loan
SP126 Automated Lib Systems
SP125 Tech Svcs Cost Studies
SP124 Barcoding of Collections
SP123 Microcomp Software Policies

SP122 End-User Search Svcs
SP121 Bibliographic Instruction
SP120 Exhibits
SP119 Catalog Maintenance Online
SP118 Unionization
SP117 Gifts & Exchange Function
SP116 Organizing for Preservation
SP115 Photocopy Services
SP114 Binding Operations
SP113 Preservation Education
SP112 Reorg of Tech and Pub Svcs
SP111 Cooperative Collection Dev
SP110 Local Cataloging Policies
SP109 Staff Training for Automation
SP108 Strategic Planning
SP107 University Archives
SP106 Electronic Mail
SP105 Nonbibliographic Dbases
SP104 Microcomputers
SP103 Asst/Assoc Dir Position
SP102 Copyright Policies
SP101 User Studies
SP100 Collection Security
SP099 Branch Libraries
SP098 Telecommunications
SP097 Building Renovation
SP096 Online Catalogs
SP095 Lib Materials Cost Studies
SP094 Fund Raising
SP093 User Instructions for Online Cats
SP092 Interlibrary Loan
SP091 Student Assistants
SP090 Integrated Lib Info Systems
SP089 Tech Svcs Cost Studies
SP088 Corporate Use of Research Libs
SP087 Collect Descript/Assessment
SP086 Professional Development
SP085 Personnel Classification Sys
SP084 Public Svcs Goals & Objectvs
SP083 Approval Plans
SP082 Document Delivery Systems
SP081 Services to the Disabled
SP080 Specialty Positions
SP079 Internships/Job Exchanges
SP078 Recruitment-Selection
SP077 Use of Small Computers
SP076 Online Biblio Search Svcs
SP075 Staff Development
SP074 Fees for Services
SP073 External User Services
SP072 Executive Review
SP071 User Surveys: Eval of Lib Svcs
SP070 Preservation Procedures
SP069 Prep Emergencies/Disasters
SP068 AACR2 Implement Studies
SP067 Affirm Action Programs
SP066 Planning Preserv of Lib Materials
SP065 Retrospective Conversion
SP064 Indirect Cost Rates
SP063 Collective Bargaining
SP062 Online Biblio Search Svcs
SP061 Status of Librarians
SP060 Lib Materials Cost Studies
SP059 Microform Collections
SP058 Goals & Objectives
SP057 Special Collections
SP056 External Communication
SP055 Internl Com/Staff & Superv Role
SP054 Internal Com/Policies & Proced
SP053 Performance Appraisal
SP052 Cost Studies & Fiscal Plan
SP051 Professional Development
SP050 Fringe Benefits
SP049 Use of Annual Reports
SP048 External Fund Raising
SP047 Automated Cataloging
SP046 Plan Future of Card Catalog
SP045 Changing Role Personnel Officer
SP044 Automated Acquisitions
SP043 Automated Circulation Sys
SP042 Resource Sharing
SP041 Collection Assessment
SP040 Skills Training
SP039 Remote Storage
SP038 Collection Dev Policies
SP037 Theft Detection & Prevent
SP036 Allocation Materials Funds
SP035 Preservation of Lib Materials
SP034 Determin Indirect Cost Rate
SP033 Intergrat Nonprint Media
SP032 Prep, Present Lib Budget
SP031 Allocation of Resources
SP030 Support Staff, Student Assts
SP029 Systems Function
SP028 Gifts & Exchange Function
SP027 Physical Access
SP026 Bibliographic Access
SP025 User Statistics and Studies
SP024 User Surveys
SP023 Grievance Policies
SP022 Private Foundations
SP021 Paraprofessionals
SP020 Managerial Technical Specialists
SP019 Staff Allocations
SP018 Staff Development
SP017 Library Instruction
SP016 Reclassification
SP015 Goals & Objectives
SP014 Performance Review
SP013 Planning Systems
SP012 Acquisition Policies
SP011 Collection Development
SP010 Leave Policies
SP009 Tenure Policies
SP008 Collective Bargaining
SP007 Personnel Class Schemes
SP006 Friends of the Lib Organization
SP005 Performance Review
SP004 Affirmative Action
SP003 A Personnel Organization
SP003 Status of Librarians
SP002 Personnel Survey (flyer only)
SP001 Organization Charts

SPEC Kit Price Information (ISSN 0160 3582)
Subscription (6 issues per year; shipping included): $200 ARL members/$270 U.S. and Canada nonmembers/$330 international customers.
Individual Kits: $35 ARL members/$45 nonmembers, plus shipping and handling.
Individual issues of the Transforming Libraries (TL) subseries: $28, plus shipping and handling.

Shipping & Handling
U.S.: UPS Ground delivery, $10 per publication.
Canada: UPS Ground delivery, $15 per publication
International and rush orders: Call or email for quote.

Payment Information
Make check or money order payable in U.S. funds to the **Association of Research Libraries,** Federal ID #52-0784198-N.
MasterCard and Visa accepted.

Send orders to: ARL Publications Distribution Center, P.O. Box 531, Annapolis Junction, MD 20701-0531
phone (301) 362-8196; fax (301) 206-9789; email <pubs@arl.org>
Order online at: <http://www.arl.org/pubscat/index.html>